SYMIE'S STORY:
EARLY MEMORIES OF AN OLD JEW
From South Haven, Michigan

By David M Liscow, MD

Foreword

Si Reznik made a habit of sharing pieces of his past with me well before he chose me to tell his story. Over the years of our relationship, he passed on a variety of bottles of aging alcohol. I had not been in South Haven, Michigan more than a few weeks in 1985, beginning as the third physician in a Family Practice group, when he welcomed me with a fifth of 1962 vintage I. W. Harper Kentucky Bourbon bottled in an elegant celebration decanter. In succession, he then gave me several bottles of liqueur and wine, some opened and some with seals intact. Among these were Forbidden Fruit Liqueur, Spanish Chablis Dalmau Hermanos &Co. and Guckenheimer Reserve. Some of these unusual vintage bottles sat on our mantle for years out of respect for the spirit in which they were given. The final bottle of wine that he gave me, he had mentioned off and on over a period of years. It was two inches of dandelion wine that his brother had made in 1924 stored in a more recent half-gallon Seagram's bottle. I sensed that this was not only the most aged of his alcohol reserves but also the most dear to him because it dated back to his childhood when his entire family was alive, including his father, Abe. He expressed his hope that I would use the dandelion wine to propose a toast at the wedding of my sons. I have yet to consume a drop from any of these bottles (my sons are still single), but I continue to value these old, stained bottles filled with their suspect contents because Si was giving me part of himself with each of these bottles. It seemed that he was trying to create a

3

bond that went beyond the traditional physician-patient relationship and was unofficially adopting me into his family. If this was his goal, he did succeed in the end. By the time I sat down with him to hear his story, I felt as if I was preparing to listen to the untold stories of my own grandfather for whom as a teenager I too infrequently provided an audience. Si temporarily inhabited the same honored recess of my heart as my grandfathers.

For fifteen years Si had entrusted his health to me. It was a natural step in his own mind when one Sunday morning while dropping off a bag of Chicago bagels, he entrusted his story to me. The time was propitious for both of us. His mind was still sharp, but it would not remain so for long. I was about to embark on the exploration of the story of past generations of my own family that I believed in all probability would intersect with his. He was a talker and storyteller by nature and I was a listener by nature and training. I did not know exactly what I would do with the material, but I intended to gather as many images as possible at the crossroads of our pasts while eliciting his stories. While he may have known about my future plans, he had chosen me because he trusted me to tell his story as he remembered it, apocryphal or not.

He had been interviewed about his early years in South Haven before, but he still resented that the sentiment of one of his fundamental stories was misrepresented. When his family came to South Haven, his father could not have bought the family farm without the generosity and trust of the original owner of the land, Bill Trowbridge. A small sum of money exchanged hands. Si felt that the original interviewer emphasized this quantity of money rather than the sentiments that made the transaction possible.

Knowing about this previous experience at the outset, I decided to respect his wish to tell his stories as his memory both saw them and felt them. My goal was not to tell the history of South Haven, but the story of Si's family in South Haven and before as Si, in his later years, remembered it.

My interests may have circumscribed the time span of his story. Without doubt Si's memory would have been clearest for the events that took place during his adult years, but I was most interested in his stories that took place during the period before 1930. Therefore we spent most of our time in what is the Never-Never land part of most of our memories, childhood and youth, and tried to bring it to life. Our story fortuitously did find a natural conclusion that coincided with the end of the period of my own interest. Our chapter of his story ends in the late 1920's when Si was not yet 20, at a time when the Reznik farm-resort era had ended and when Si, after tasting the bitter sweets of city life, has returned to South Haven to begin his adult life.

We sat down many weekend afternoons in his front living room with picture windows giving a view of lake Michigan. Si's wife, Belle, who probably had heard his stories a "few" times before, discreetly entered the room and offered us tea, and then went about her housework. We usually had no agenda other than to populate the past with his stories. While his stories emerged, I took notes much like I do when I listen to a patient's story during an office visit. Occasionally, I stopped him to get more details or to try and place the event on a timeline. After each session, I returned home and before the stories grew stale, I rendered the notes into nascent stories and wrote down any follow up questions. We repeated this process many times until very few new stories or details from this chapter of his life arose. Our meetings eventually came to a natural conclusion.

I assumed that the stories that Si recalled from his childhood survived because they either had entertainment value or formed the building blocks of who he believed he was. Some were the furniture of his life and others were the cornerstones, but they were the ones he carried with him well into the ninth decade of his life. Besides arranging these stories organically and coherently, I endeavored to enhance their immediacy without altering their plots or meaning to Si. Living in South Haven for 15 years and experiencing decades later many of the same areas of Southwestern Michigan that provided the setting for Si's stories gave me a cache of images that I used for this purpose. I also did a minor amount of research, particularly, regarding the circumstances of the Jews in Russia during the last half of the 19th century and early 20th century. I used it minimally when providing background for Si's memories of his father, Abe, in Russia before he immigrated to the United States. However, the understanding of Jewish life in the small villages of Russia became more important when trying to understand the decisions Abe made when he lived in the United States. I believe that this process of telling and arranging the stories allowed Si to live in this precious remote period one more time with his long dead father and less so, with his mother, brothers, and sisters. It allowed him to develop a deeper understanding for the people who were responsible for standing him on his feet.

There were times when I may have had reason to doubt some of the details of Si's version of an event. Distinguishing between tales and true stories was not my object; to illuminate the world around Si's childhood as he remembered it was my object. So whether the eye of Sanderovich, the shoemaker, really hung from its socket after a fight about the placement of a synagogue was not as important as Si's belief that it did and that he assisted him in receiving appropriate medical care. I have no doubt that some other present-day octogenarian living in South Haven who experienced the same events might contradict, or a least, change Si's version of the events. This possibility no more undermines the significance of Si's foundational stories than it would some of the foundational stories of any person, historical international icon, or for that matter, world religion.

Ultimately, Si would determine whether I was successful in bringing his early memories to life in print. Did he visit that distant world that Symie, his childhood self, inhabited? Did it resonate harmoniously or discordantly with the voice he wanted to leave in his advanced years? Si did have an opportunity to live with a nearly finished version of our manuscript before matters of health began to dominate his days. I recall few details of what he had to say. I remember only that he could not read it without crying.

That Si was moved to tears presumably of recognition and the sweet sorrow of nostalgia would have brought the project to a conclusion had I not known that Si wanted to publish the unfinished manuscript and had I not believed that others might find Symie's world as absorbing as I did. Not only was there an unfamiliar story about days gone by, but also there was a more familiar love story about the hero of his childhood, his father. Without yielding to the technology of the 20th century, or the 19th century—he left that to his children—his father managed to give his children everything they needed to prosper in their new world, the United States. His father is the towering figure of Si's childhood. For that reason, of all the items that Si gave me, I cherish most a pocked and stained wooden level that his father made. The level dated to an interlude in England around 1909 when his father had left Eastern Europe and was desperately poor trying to earn enough money making furniture for passage to the United States. The level crossed the ocean with him. I can only hope that were this level laid across *Symie's Story* its bubble would sway as if to a distant favorite tune until the end when it would come to rest on the meridian.

Contents

About 1920

Dave   Lou,   Joe   Si Meyer   Charlie
Levitt   Cohen,   Reznik   Reznik Reznik Peirce

Kate Reznik Ann Reznik   Pa Reznik   Ma Reznik   Fannie Reznik
Levitt   Cohen   Abraham   Hinde   Peirce

                          Holding   Phyllis Levitt
                          Seymour   on Lap.
                          Cohen

10

Introduction

He was more than twice my middle-aged years, and as I listened to him recount the stories of his remote childhood, I could not help but ponder the origins and nature of memory. When God created the human memory, He dipped a thimble into His own memory and gave it to humankind. He gave it qualities that belong to a starless night and the clarity of day, that resonate with a granite boulder and a mushroom, sturdy as steel and vulnerable as crystal. Those precious gems we hoped to retain, memory allows to slip away; and yet, it clings to those painful barbs we wished it would purge. We are annoyed by its clumsy and self-serving attempts to record events, but we marvel at its ability to suffuse events with the warm and endearing shades of love or the endless detail of passion. What it treasured in a moment of revelation seems rubbish in a sober moment of reason. A single human memory, glimpsing human existence for a moment no longer than a lifetime, is a flash of light in a tiny dark cubicle compared to God's concentric, radiant, eternal beams.

But herein lies the beauty of the human memory, the wellspring of the story, the notion that any individual's memory potentially records a unique scene on the frieze of human experience. We thirst for the images of our past in the memories and stories of others, because somewhere in them lie our roots, a connection to something larger and longer than our meager individual existence. Among the memories out there, those that have drawn a long taper amaze us. Like the lights of Hanukkah, they shine brightly for what seems a miraculous amount of time, shedding light with the spoken word on the events of generations back. This is no casual feat for these individuals must reach beyond the octogenarian decade, with their memories intact. They must retain the zest to tell their stories, and the courage to find an interested ear to record them.

When such an individual sits down to tell his story of "Once a long time ago", what should we expect? Can he remember the scent, the exact word, the exact intonation? Can he give the objective point of view on the events of his day? That should not be our expectation. The memory is a receptacle with entrances chiseled by fear, love, hate, joy and self-preservation, just to name a few tools. It develops leaks at the bottom, created by these same tools, and by lack of use, satiety, and the exhaustion of age. The memories that do not escape are the seminal events, the ones that made us who we are, our heroes like national monuments and our values like the Constitution.

One's memories change state (when they find words), much like a gas changes to a liquid or a liquid to a solid. They flow into a unique form sculpted by its storyteller. And when he forms the story again, they take similar shapes, but with finer details, ones that might be more appealing to the senses of the beholder - still the truth, but a more honed form of it. Perhaps some details are added by the recollections of others, from other historical accounts, or even from later experiences that awaken or amplify old memories, but one's stories do change.

I had the privilege of sitting down with Si Reznik, a man well into his ninth decade of life, who remembered the stories of his childhood as clearly as he remembered the events of the day before. When his eyes stared into a scene from the past, I knew he did not see the lake through his front picture window, but rather some event from a past as distant and as beautiful as Mount McKinley on a clear day. He said, "I can see it." What delight there was in visiting the events of another lifetime, and the loved ones who populated that time, the living memories associated with the pictures in old photo albums, infrequently opened. At times he seemed to be narrating the newsreels of his memory, edited not only by his advanced years and his generous and somewhat limited human perspective, but also by the humility and wisdom offered by nearly nine decades of life.

Si Reznik took me through the chambers of his early memories and dusted off the relics, revealing their nature to me as honestly as he could. Some events he obviously had witnessed and remembered acutely enough that his eyes watered from the acrid scents. They filled his octogenarian spirit like a flaccid windsock taken by the zephyrs of spring. He remembered some details of his family's life from decades before his own birth in 1912. The details of these events were obviously passed down to him just as he was passing to me the events that he directly witnessed. The details are sketchier than the ones he directly witnessed or experienced, but they served, nevertheless, as a background for the events which took place after his birth. Many of the memories poured out as undated stories, having occurred some time during a five-year window. Only after considering the story in the context of everything else he had recounted were we able to narrow the period of the event to a couple of years. Fundamental events, on the other hand, like the death of his father or the year the first South Haven synagogue was built, he remembered with more clarity.

Most people would find it remarkable how much this simple, hardworking, salt-of-the-earth man remembered about his childhood years. At half his age, many of us could not recollect a fraction of what he was able to describe. To us, our childhoods seem generic, with the same hoops and hurdles that other children had to jump. We traverse the memories of our childhood as we would a neighborhood filled with track houses spit off the same mold, our particularly heroic or devastating moments showing up as the only differences, like ornamental trees or shrubs. Our memories of childhood lack the piquancy to generate the energy to retrieve them from storage. Our childhoods were a passage, a means to an end, a pre-conscious state during a period when remembering was not one of our responsibilities.

Simon Reznik remembered his past, not only because he has an unusual memory, but because he perceived the events of his life as unique and the people in it as heroes. He was the youngest of five children. While he did not shirk any responsibilities given, as the youngest, he was given fewer of them and was free to make observations, store them and cherish them. His childhood was filled with people he admired and loved, from his parents to his siblings to his rabbis. I do not mean to suggest that life was easy in the Reznik household, because it was not. Life required survival from season to season; it overflowed with hard work and drudgery, and not an ounce of privilege. A life filled with the potential of so many failures overcome only by human endurance, sweat, and courage, offered an unending succession of successes as well as worries. Such circumstances generated a sense that what he was experiencing was unique, that his family was among the pioneers of the country around South Haven. I believe that, at least in part, this is why the boy, Symie, now grown to the elderly man, Si, recalls large tracts of the odyssey of his childhood.

Before America

Si recalls only a few treasured nuggets from the past of his father, before he was born and his father left Eastern Europe in the late 1890's. These stories evoke the circumstances of many of the Jews that began life insulated in a shtetl, a small Jewish village, in Russia, always at risk of falling victim to the scythe of the unpredictable hatred and ignorance of pogroms. Avram Reznik, Si's father, was born in 1871 in the Vilna Province of Russia at the crossroads of what is now Poland, Lithuania and Russia. Had he been born a century earlier, he would have been born in a shtetl that still had a vague memory of independence from its larger neighbors of Poland and Russia. A weak monarchy of Catholic Poland would have claimed his shtetl as part of its territory, but would have left much of the governing to a local Jewish council, leaving the Jewish way of life largely undisturbed. To the north and east, the Russian empire would have been flexing its muscle, but it would still be a couple of decades before Catherine the Great would carve off the Polecized Vilna province, a morsel for her bottomless land hunger. After Vilna fell into Russian hands, a series of decrees by the Russian Czars stripped the Jews of everything but their history, their religion, and their way of life. While they could farm, they could no longer own the land which they tilled. Ancestral lands under Jewish ownership for centuries were first turned over to non-Jewish Russian magnates. Later, after the reforms of the more benevolent Tzar Alexander II, what remained was divided among the poor Russian serfs. A Jewish male born in 1871 in the Vilna Province of Russia, therefore, had no prospect of owning land. What's more, he faced the prospect of forced conscription in the Russian military for a minimum of five years. Five years was a merciful decrease, though, from the twenty-five year term of forced conscription a decade earlier under Tzar Nicholas I, a term that often began after a boy was kidnapped before his twelfth birthday.

Nevertheless, the Jewish way of life in the shtetl, ingrained over the centuries, survived an assault that at times was worse than the tenth plague God visited upon the biblical Egyptians. The shul, or synagogue, remained at the center of any Jewish boy's life. Not only did it serve to provide religious education and camaraderie, but also it provided a reliable haven from the turbulent, changing events of the outside world. It was a world fortified by more than a thousand years of Jewish tradition under the wing of God. While it may seem that the shul was an escape from an inescapable horrible reality, this world was their inheritance, their chest of treasures, as God's chosen people. When the time came to leave shtetl, life as oppression degenerated to brutal destruction en masse, the central role of the shul in shtetl life would be deeply missed in the new world of opportunity. For some, possibly for Abraham Reznik, the establishment of a shul that served the same central role as it did in the shtetl of the old country was akin to a search for the Holy Grail.

Abraham Reznik enlisted in the Russian army before his eighteenth birthday. When he recounted his experience in the Russian military to his son in later years, he did so with pride and with a fondness that belied the horror stories of other conscripted Jews. In contrast to the general trend of Jewish harassment and constant pressure to convert, he was given time each day to fulfill his daily Jewish mitzvoth, or ritualistic good deeds. He arose at four a.m. each morning. He had one hour to don his teffilin, two small black boxes with black straps that he bound to himself daily to affirm his devotion to God, and another to groom and feed his horse. His commanding officer must have thought, "What harm can there be in this peculiar custom of making strange incantations while wearing a couple of tiny leather boxes. I've heard that a little piece of parchment in the boxes has nothing but gibberish on it. So what if it falls in the hands of the enemy? As long as he does it on his own time! And if he strangles himself on one of those long straps, where is the loss as long as he takes good care of his horse?" His commanding officer impressed upon him that the wellbeing of his horse was more important than his own, so the horse was always fed first and starved last. Abraham accepted this as a matter of necessity in the military, rather than a punishment for his Judaism. Unwittingly, the commanding officer had given Abraham permission to erect the fortifications, which girded him from such insults to human dignity. Could he have known that each of the seven times Abraham encircled his wrist with that leather strap, he was affirming his loyalty to God Who freed him from a bondage much worse than the one the officer represented?

His commanding officer would also allow him to attend shul whenever they passed through a part of Russia where there was a large enough Jewish population to support one. The majority of the Jewish population was confined by decree to western Russia in occupied Poland and Lithuania. This freedom would some day change his life forever.

During Abraham's tenure in the Russian army, the Russian empire, with powerful and friendly countries to the west, looked to the east. In preparation for an assault or defense of this area, perhaps from Japan, the strongest country to the east, Russia began the arduous task of establishing rail transportation to the area, constructing the Trans-Siberian Railway. By 1901, this railway reduced the journey from Moscow to Vladivostok, Russia's largest city on the Pacific, from eighteen months to two weeks. Prior to the construction of the railroad, the troops had to trek over a laborious series of land and water routes. Abraham Reznik probably participated in this push to the east. As there was no off-season in military construction, Abraham spent the winter months marching toward Japan, constructing a railroad across the thick ice of a frozen lake rather than taking the lengthier route around it. He started the long trek home well before he reached the goal destination, the Pacific.

Perhaps because of his experience in the army or in spite of it, Abraham Reznik developed an air of confidence in those days. Nothing was impossible. He re-entered the familiar territory of the Pale, the area of Russia to which the Jews were restricted, at the end of enlistment; he had the strength and bravado of a seasoned soldier. Passing through the various towns and villages, he was able to attend shul again. One spring morning attired in his soldier's uniform with a fresh shine on his boots, he was leaving shul after an overnight rain had left the streets sloppy and muddy. Abraham noticed a petite, pretty girl come to a halt before a particularly deep mud puddle. No time for a matchmaker, his instincts took over. He marched to her side and gallantly scooped her off her feet with an iron arm around her waist and planted her on dry ground on the other side of the puddle. Then he properly introduced himself to her. With her father's permission, he began to court her. That is how a long trek home, the draw of the shul, and a spring mud puddle conspired to bring Abraham together with his future wife, Hinda.

Hinda Swimmer was one of nine children living in a one room thatched cabin. Her father, Zelig, had no land of his own and worked as milliner at a local estate of a Russian magnate. Though poor, Zelig was a charitable and pious Jew. He worked at the
end of a conveyer, bagging cracked wheat, oats or rye. He knew that the magnate would never sully himself and inspect the guts of the process by which he made his wealth. He need not watch because God was watching poor Zelig. God watched Zelig, but Zelig knew of stomachs more poorly supplied than his own, and he inconspicuously squirreled away small portions of cracked grain. Over the objections of his less charitable wife, he distributed the grain to his widowed neighbors or heated it up in water and fed it to a devoted cadre of cats and dogs. As he scooped his tiny piles of grain, he soothed his conscience and made peace with God, repeating in Yiddish over and over, "God forgive me." God must have forgiven him, because he was never punished for these transgressions.

Zelig did suffer at the hand of his countryman though. He watched his children leave his shtetl one by one for the United States for a life free of persecution. They left either as they came of age for the Russian army or as the pogroms, the pillaging of Jewish villages, rattled neighboring villages. Many of the Swimmer family eventually made it to Canton, Ohio. Zelig remained with the village and the way of life he knew. He watched the cruelty around him progress from property destruction to full-blown pogroms. Jews were slaughtered like cattle for gentile consumption. They died with terror and God in their hearts for no apparent purpose. Rather than endure the fate of a fatted cow when the pogroms were at his doorstep, Zelig took to the woods in the middle of winter.

Si remembers the day when a letter arrived for his mother from one of her brothers in Canton, Ohio. At the time, Si was a young child. While he had seen his mother exhausted from incessant chores, he had never seen his mother cry. He had never seen her wear the black shawl that hid the anguish of her sobs. The letter from her brother described how her father had fled from Vilna into the woods and that until recently, there had been no word of his whereabouts. He had been found, frozen solid. Among the icicles, adhering to his long beard and mustache were chewed crumbs of bark. Si's mother mourned pitifully for months. She mourned while the Great War raged in Europe and the Russians purged their land of Jewish countryman. Ironically, they killed only those Jews who were not serving their country on the western front, defending their homeland against the Germans.

With his pregnant wife, Abraham left Russia long before Hinda's father suffered his final fate. During the pogroms of 1896, they left the land where their forefathers had lived for centuries. They left in a hurry, with few possessions and a pittance to fund the expedition. Though bound for America, by the time they had purchased fare to cross the Baltic and the North Seas, and finally the English Channel, their funds had completely evaporated. The longest leg of their journey remained, steam passage across the Atlantic to Ellis Island. Meanwhile, they lived among many Russian immigrants in a more tolerant city, London. They had no skills, only a rudimentary understanding of the English language, and a child on the way. While Abraham had traveled the undeveloped world of the

Russian frontier in the army and was a man of the world in this limited way, living in a foreign industrialized country presented a new and frightening set of circumstances. But poverty was no stranger. They had a few articles of clothing beyond the clothes on their backs. Experienced in the art of hardship in the Russian army and raised in a country where land ownership was a privilege only for others, beginning again with nothing, making do without, was nothing new.

Abraham found work in a cabinet shop. He began as an unskilled laborer, but over the one or two-year period required to earn his Pekla, his passage, to America, he learned to skillfully manipulate the tools of the trade. By the time he left, he had made a set of tools to travel the Atlantic with him: a level, a set of vices with hand carved wooden screws, and a plane. While handling tools with the confidence of an expert brought a measure of self-sufficiency, he did not forget that woodworking was a step along the way to America. They lived a meager existence, cramped in an overcrowded part of London that was teeming with street life. Abraham left and returned in the darkness, working 18-hour days. One day, upon returning home from one of these long days, he learned that he was now responsible for earning Pekla for three. Their first son, Joseph, was born safely at home.

A Long Layover in Sandwich, Illinois

After successfully walking the welcoming gauntlet of Ellis Island, they made their way to Canton, Ohio, and stayed briefly with other immigrant members of Hinda's family. From the small industrial center in Ohio, they moved to the more promising undeveloped frontier of farmlands in Dekalb County, Illinois. They settled in the small town of Sandwich, Illinois, located forty miles west-southwest of Chicago and twenty miles from Aurora, Illinois. Here, Abraham, now a sturdy, hardened, field-tested man in his late twenties, was prepared once more to see what a bounty of hard work and faith, with just enough sleep and procreation to get by, could make from nothing. The soil was black and fertile, rich from millennia of decaying prairie grasses, but still, he could only work it, not own it. He could not share in the American Dream. In Dekalb County, his family settled into a sharecroppers' life, an existence that permitted subsistence only. While it offered safety and freedom of religion, it did not bestow financial independence or the pride of ownership. They had the freedom to practice their religion, but not a community with which to share it. As the Reznik family grew from three to eight, finding a Rabbi for a circumcision or a baby naming was a major undertaking, involving either hours or a day of family travel across rugged terrain or a dedicated frontier Rabbi willing to make the journey.

When the Reznik family moved to the Sandwich farm, they occupied an abandoned, dilapidated hovel that had been abused by the previous tenants. The front of the house was missing only a "condemned" sign. It consisted of a set of rotted steps and an askew termite-infested porch, a strong gust short of collapse, and gray weather-beaten clapboard – some boards dangling, others spongy with rot - forming a sieve to the howling winter winds. Abraham fixed enough to keep out the weather, but for the most part, the family entered from the stairs in the rear and squeezed their life into the back of the house. The house was too far-gone to ever be a home, which they would willingly share with company.

Abe trudged through three legs of work every day. First he tended to the tasks of a small dairy farm that required his strength and experience. He spent most of his time with the crops, mainly corn, which was most labor-intensive during planting and harvesting. Then he hitched his team of horses to his wagon and set off on the rutted, turn-of-the-century dirt roads to collect the day's milk production from the neighboring farms. Swatting horseflies on a mild June day under a blazing sun with gentle breezes stirring the dust augured a good day. Watching a dark eruption of clouds on the western horizon as a gusty front peppered his eyes and nostrils with dust and grit had the making of a difficult day. Whatever nature had in store for him, when he finished collecting the milk and unloaded it at the train station, he returned home for dinner with his family. After this brief respite, he left for the third leg of his workday, unloading coal cars for twenty-five cents an hour. By the time he completed this job, laboring always as if God was watching, plastered with soot from the soft coal that worked deep into his ears, caked his nose hairs and seemed to find every opening in his clothing, only four hours separated him from the next day's cycle of work. Each night before retiring he sponged himself off with a basin of fresh cool water. He hardly noticed anymore that his fair, sunburned, Eastern European complexion stained by dried streaks of sweat had been transformed into the black earth itself, cut unevenly with swaths of bleached white roads.

The Sabbath was a solemn day of well-deserved rest, but it came at a price when financial reserves were meager. The Rezniks hired a gentile neighbor to tend to the farm's daily chores that did not heed their cycle of creation. The day allowed Abe to spend a rare stretch of time with his growing family. The younger children who hungered for their phantom patriarch during the week feasted on his constant stolid presence on the Sabbath. In the close quarters of the back of the house, Abraham and the smoke of his Turkish tobacco transformed this little cloistered world. Not in haste, but as if he had all day, he studied how a week of growth had changed them, a new Yiddish word or a pinch more of confidence toddling around the room. Though forbidden on the Sabbath to fill a basin of water, he could impart a soldier's past and fill their heads with stories of the Judaism of his ancestors. Gazing upon his brood, he recognized with a consuming serenity that an army would not steal his sons from their mother and that his daughters would not have to suffer the abyss of early widowhood and fatherless children.

Hinda, the petite, stoic Reznik matriarch who had so far not realized the benefit of making the journey across the vast Atlantic, kept the expanding household going. While in Sandwich, the Reznik clan grew from three to eight. Si brought up the rear in 1911 and his three sisters, Anna, Kate and Fanny, and his brother, Meyer, filling in the middle. As Joseph, the oldest son, approached the age of Bar Mitzvah, he became a more helpful hand to his father on the farm. To the other children still under foot, Hinda allotted small chores such as feeding the chickens or cleaning the chicken coop when their little bodies and growing minds grasped the whole of a little task. In most cases, the instruction took more time than the small task itself, but Hinda knew that once a routine was established, it would be done silently, without reminding.

When the children came of age, one by one, she saw them off to a clapboard one-room schoolhouse that served their farming community. Still shy with her own English, she marveled and recoiled simultaneously as her children's skill with English surpassed her own. Overcoming some of her reticence with the language of America and a niggling tendency to feel antiquated, she tried to welcome the new words that her enthusiastic brood shared with her. For fun, she would bark out orders in these new words when the chores had to be done.

Mostly she struggled with life in America, a life she knew she had to accept or she would remain miserable. On the surface of things, she was living on a farm that functioned like it would anywhere else in the world, filled with unpredictability and toil; these conditions were familiar and in some ways comforting. But she had to accept that she was sending her children off to a school, not a shul, where English, not Yiddish, Hebrew or Russian was spoken, where the history of a foreign country was taught, not the Torah and the Talmud. Most days, the daily drudgery was too great to recognize that she was taking part in the history of that foreign country.

Besides raising her children in a land to which they were assimilating more rapidly than she was, she was faced with an endless track of chores. Had she complaining hands and arms, she would have gotten nothing done, because everything she did required a dose of elbow grease. From the crow of the rooster until the silence of sleeping children, her hands boarded a milk train, stopping at a variety of tiresome stations: first stop – kneading ripe utters; next stop – grasping the soiled canvas of her husband's clothes, dunking her hands dozens of times in a basin of increasingly murky water as she freed the fabric's embedded dirt, rubbing them against the washboard; next stop – hanging clothes to dry; next stop – darning overalls, socks and anything else that had a scrap of life left in it; next stop – scrubbing the uneven boards of wooden counters and floors; and multiple stops through the day to prepare meals of fish, cheese, potatoes and bread. By the time she was done rinsing the plates, cups and utensils from the evening meal and making sure the table and counter were crumb-free to avoid pests, and by the time she lead each of her children through the sequence of rituals that brought the soundest night of sleep, the day clocked in chores rather than minutes and hours, mercifully concluded. Even her anxieties about her children and a husband who worked too hard did not detain sleep.

The only sure thing in the Reznik's life in Sandwich, Illinois, seemed to be the enduring relationship between hard work and food on the table. Only economic depression or personal illness could upset this fact of their tenuous existence. One spring, before the year's planting took place, after the risk of another frost passed and the field finally dried enough to plow, a contagious illness swept through the Reznik household. The entire family was rashed-up, feverish, weak as starved animals, disdainful of food and consumed with a desire to sleep endlessly. They would lose none of their children to the illness, but there was little that could be done, other than adhere to a strict quarantine. The coal soot, Abe could forgo temporarily.

His neighbor's milk – there were others who could collect it. Milking the cows and feeding the animals – as long as the family adhered to the strict quarantine, their neighbors, most of whom had been in the same position, could send their more expendable, youthful hands to take turns pitching in. But for the spring planting, the window of opportunity was drawing to a close. If this cycle were interrupted, there would be no winter store of food for beast or human and there would be none to sell to fund other winter purchases. Each passing day added a week to Abe's burden of worry. He was lost at sea and drowning, cut off from his lifeline of hard work. His fear focused him only on the possibility that the water might engulf him and blinded him to the possibility that a rescue raft floated a few swells away. Even his prayers were lost in the swells that lapped at his ears.

As the prospect for a spring planting grew as dim as the predawn light that entered the windows of their cell, an awful racket awakened the Reznik family. Never before in this peaceful community had they heard such a din! Perhaps the authorities were taking the next step after quarantine before they could awaken and dress – drive out the inhabitants, plow the diseased quarters into the ground, then make a big bonfire as an offering to the gods of disease. Or did it awaken memories from the Pale in the elder Rezniks that it was a mob carrying torches and axes? The Rezniks did not know whether to huddle in the center of the room or peer through the closest window. Gradually, as the invasion moved closer and his ears adjusted to it like eyes do to furniture in a near pitch-black room, Abe could make out some of the commotion. He recognized the hollers of his neighbors, the whinnies of horses, the revolving of iron-rimmed wheels over uneven turf, and the faint scraping sounds of plowing discs on untended turf. These familiar sounds did not usually add up to a terrifying racket; they were usually the peaceful sounds of hope. He moved toward the window. In the gray light of the early dawn, assembled outside, he could make out what appeared to be every piece of farm equipment in a five-mile radius. In that one day, from the terrible dawn to the last wink of dusk, the Reznik neighbors turned over all the soil and sewed a season of seeds in the manner of a barn-raising. Abe might have wept with thanks and jubilation had it not been tainted with the guilt that others were doing his work. He would not go outside to thank them. God would see to it that he had a chance to thank them in kind.

In his later years, Si Reznik would remember this story, passed down to him during the tedious hours of goose feather plucking. It was part of the family's defining lore. The elder Rezniks who were all too familiar with hatred from their gentile neighbors and who were accustomed to viewing every gentile with suspicion regardless, discovered that harmony was possible between Jews and Gentiles. The post-holocaust Si still believes that had Jews and Gentiles lived together as they did in Sandwich, Illinois, there never would have been a Hitler.

Lake Breezes Fan Abe's Ambition

Sandwich, Illinois, still was not capable of fulfilling Abe Reznik's ambition; even harmonious relations with gentiles did not satisfy his most fundamental needs. In truth, any type of relationship among Jews was preferable to peaceful relations with gentiles. And he still did not have a plot of land to call his own. Between 1911 and 1912, with the assistance of the Jewish Agricultural Aid Society (JAAS), the possibility of achieving both of these goals developed. The JAAS was a philanthropic organization centered in Chicago. Its mission was to help Jews buy and establish their own farms. While many European Jews aspired to become professionals, educators, businessmen and financiers, this organization recognized that Jews had a long tradition in farming in the shtetls of Eastern Europe and that farming, for a group of Jews that were still adrift and struggling, offered a means of escaping a suffocating city existence for potential prosperity in rural America. By offering low interest loans, the society provided the means for Jews to resettle in rural areas of the mid-west, at times in large enough numbers to form a Jewish community. For some of the first generation Jews who originally settled in rural areas sparsely populated with Jews, this resettlement had the potential to realize some of the sorely missed traditions of the old country.

The Agricultural Society directed Abe Reznik to a vacated lot of land three miles to the east of a small town called South Haven, located on the southeastern shore of Lake Michigan. They might have sold Abe on the area with the following spiel: "It's an old lumber town that Chicagoans now use as a vacation spot because it has good beaches, good fishing and a good harbor. The winters are snowy – worse than on this side of the lake – but the cold is not as harsh because of the warmer lake breezes in the winter. In the summer it's not so hot because of the cooling lake breezes. Lake Michigan gives it an ideal growing season for a place so far north. The soil's not as rich as what you have here, but the climate makes up for it. Great for fruit trees – apples, pears, cherries. You won't find many Jews right in the area, but you will find some to the south twenty or twenty-five miles and to the east about forty, really too far to be practical. But you could be the start of something, the crystal of a vital Jewish community; you and a couple other families; all it would take. Lots of farmland in the area for sale. Might be rough at first, but can't be much worse than what you have right now".

Abe made the trip to South Haven to examine the forty-acre plot of land. He found a two-story house that had been abandoned for a few years and had fallen into mild disrepair and at one time had functional indoor plumbing. A wood-burning stove located in the kitchen adjacent to the master bedroom heated the first floor. There were several more unheated bedrooms on the second floor. Overall, he found the house to be a sound structure and a vast improvement over the reclamation project in Sandwich, Illinois. It might serve as his final home.

The previous occupant had raised racehorses and kept them to the west of the house in a large barn filled with sturdily constructed stalls. Another two-story barn, for poultry and dairy cows, lay northwest of it. The paths between outbuildings, at one time well worn, were beginning to fill in with goldenrod, sassafras shoots, and other weed grasses. On slightly higher ground some twenty-five acres of the spread had been cleared and was tillable. Fifteen acres on the northwest corner, part of the lowlands of Trowbridge Flats, was still wooded and was in the flood plain of the Black River that skirted it to the west and south. Perched on the edge of the cleared land overlooking the flatland stood a small, still productive vineyard with a variety of grapevines clinging to rusted wires strung on posts, leaning willy-nilly from neglect. Abe considered, "Just a little attention here might ensure a year's worth of wine. Certainly would be an oddity in the corn belt of Illinois". Also in need of attention, separated from the house by the rutted drive entering the farm, was a small orchard of fruit trees, mainly pears, apples, and cherries. This farm and the land across the road to the north were owned by Bill Trowbridge, a local constable with a reputation both for boozing and for decency.

Bill Trowbridge's father, a trapper, had settled this area around the Black River and had given it his name, Trowbridge Flats. His son, Bill, had no qualms selling part of it to someone who would be a decent, hardworking neighbor; he was not looking for another member for his church either. Upon walking the land and reworking it in his imagination, Abraham Reznik decided to move his family. He only needed to strike a deal with this easy-going stranger. He assured Bill Trowbridge that he was an experienced farmer and that he would make something of the land, but he didn't have much money to buy it. Unconcerned, Bill responded, "Do you have a right hand"? Confused, Abe responded that he still had his right hand. The trapper's son then offered his right hand to Abe saying, "Then we have a deal". Abe took the hand of the man who never would give him any reason to doubt his sincerity. They settled on a down payment in the neighborhood of fifty dollars. Bill Trowbridge reassured his still incredulous buyer, "Don't worry about the rest".

Pioneers Head East on a Milk Train

The journey from Sandwich, Illinois, to South Haven in 1912 was a cornerstone in the Reznik's mosaic of hardships. The trip was planned to save every penny for the next phase of their life. Any comfort taken was a necessity. Hinda Reznik with her three daughters and her infant son occupied a single set of facing benches on the passenger car for the duration, a day and a night. In an open-slatted boxcar, Abe and his older sons shared their journey with the livestock that they hoped would be fruitful and multiply in South Haven. The uneasy cattle slipped and, at times, flopped, jostled by a train paying heed to every tiny station between Sandwich and South Haven. The Reznik boys figured that they saved a cow more than once from being trampled by its neighbors when they helped a fallen animal back to its feet. Sleep for the shepherding Rezniks occurred when the train achieved a steady speed long enough that the cattle relaxed and stopped their lowing complaints. During an occasional longer stop, Abe took the boys inside the train station so they could steady their legs on the unmoving earth. There they purged the stench from their lungs with fresh air and used the indoor plumbing to relieve themselves and wash their hands. Ravenously, they ate some of the black bread and farm cheese Hinda had packed. Had they been a self-conscious lot, unconsumed by their mission, they might have avoided populated, well-lit public places, mingling with delicate noses and refined eyes unaccustomed to the residue of life among cattle. The Rezniks used the less public stretches to fertilize the countryside with cow piles mashed into the floor of the cattle car. For the Reznik boys, it seemed that the journey would never end. Each stop was followed only by another one that was not South Haven. They stopped asking and resigned themselves to more of the same, the pitched black of night or the broken light of day, the metallic droning of the train wheels, and an increasingly weary eye for a fallen cow. For the elder Reznik who had experienced steerage across a vaster endlessness for a land with even greater surprises and expectations, the well of patience had been drilled deeper.

The stops finally came to an end as the train pulled in alongside the Black River. They shared their arrival with visitors from the west side of Chicago who had made their journey in the luxury of a steamship. While someone catered to the needs of these carefree vacationers familiar with leisure, the Rezniks gathered their bags of clothes and trunks of utensils and put them on the back of the wagon. The children gawked at the figure-eight roller coaster and Chautaqua tent across the river. They watched giggling girls disembarking the Steamship South Haven as the more settled vacationers strolled south on Williams Street in petticoats and parasols. To the east, a well-manicured gentleman departed the Pierce-Williams Basket Factory just before workers filed out with stained, calloused fingers. Barely mindful of the life of a resort and weary of sustenance labor, Abraham hitched the wagon to the team of horses that made the journey from Sandwich. Then he and the older boys unloaded the cattle. Had they been sentient creatures, they would have expected a fate much worse than a three-mile trek to their new pasture. The Reznik boys took a few minutes to cleanse their lungs with the Lake Michigan breezes and absorb the sweet scent of the confection booth intermingled with the smell of fresh fish. Their exhaustion from the long journey began to ease as the expectation of arriving at the portal of their future asserted itself. Their father thought, this isn't Ellis Island, but it is a new beginning. The moment passed, and they united in the task of herding the relieved livestock up Williams Street, turning eastbound on Phoenix for a straight three-mile shot to home.

Abraham Reznik was 41 years old now. The years had put an ache each morning in his laboring joints and the Turkish tobacco had shriveled his seemingly boundless stamina, but time's toll was offset by a rejuvenated sense of purpose. If he were to take stock now, fourteen years since he left the land of his forefathers, he would have noted that he had made some gains, but at a sacrifice. Without handing him anything material, America had given him enough security to follow God's commandment to be fruitful and multiply. It had whittled away the incessant dread that he was committing his children to a life of misery and persecution. It had given him hope without dashing it with insidious murmurs of hatred and threats of annihilation. Up until now he had little to show for it, but he had been given the opportunity to work, as hard as his body would permit, toward a better life. America had offered him an entirely different reality, freeing him from the conceptual yoke that he was born in a world in which he was preordained to experience joy from his religion as long as he suffered its accompanying persecution. How absurd it would have been in Russia to hope for an organization which would have had enough power to resettle him, a Jew, so that he might own land! Yet, as bad as life had been, as deadly as the future had loomed since he had left it fourteen years ago, he still had not remotely realized the communal religious life of the shtetl. Had the time come within this new land, three miles outside of a resort town, when it was no longer an empty prayer that he might begin to recapture an inkling of that ancient tradition?

The Farm Becomes Home

What were the possibilities? What would it take?
He was planting roots three miles from a thriving resort
town. Daily during the summer, a steamship, the *City of
South Haven*, shuttled a village of well-off passengers
from Chicago. Among this hoard there were certainly
some Jews who had earned a measure of leisure. While
their religious focus was most likely tied to their shuls
at home, their need to practice their religion did not
disappear when they were not working. If South Haven
had to welcome any visitor with open arms, would it
not eventually develop a level of tolerance for all
newcomers? Would not South Haven learn to keep
their gates open even wider than those at Ellis Island?
Hopefully, a hardworking newcomer, Jewish and still
poor, could put on a clean pair of trousers and drive his
wagon into town and, in his thick Yiddish accent, ask for
those few things, which the farm and common sense
could not produce. If one Jew could get the help he
needs form M. V. Selkirk's Carriage and Harness Shop, E.
M. Gale's Flour and Feed Store, Merson Brothers'
Grocery Store or Hale's Clothing Store, and if he were to
share his experience with others, wouldn't other Jews
want to follow? But if Abe Reznik had these thoughts in
1912, he had them silently in the transitional moments
before sound sleep. There was an unbroken chain of
tasks to accomplish in revitalizing an abandoned farm.
Though it was spring, he knew that parsimonious
winter was always around the bend.
    The instinct of a man who knew small farm
husbandry took over. There could be plenty of dairy
products – fresh milk and cheese with a little know-how
and patience, and butter with a little elbow grease. The

small purse of money he had accumulated when he sold his chickens and geese in Sandwich, he would sink right back into the egg-layers and breeding geese. Joe Reznik, already revealing himself to be a master handyman around the farm at the age of 15, was given the responsibility of constructing a poultry coop southeast of the cattle barn. For the hens, he rigged up crates covered with heaps of straw to gather guano and collect eggs. When the coop was completed, he reinforced the stalls for the cows in the dairy barn and replaced rotted boards. Tzalie (Abe) made plans for the tillable land, designating some of it for planting and the rest of it for foraging for the cows and horses. He had to nurture his feed crops in grayish-brown dirt that was once the floor of a dense Michigan forest of oak, beech, maple, poplar and hemlock. The soil was sandy and not as deep and rich with organic life as he had grown accustomed, but it had lain fallow for a few years. Years of grass and weed overgrowth protected the earth from erosion and restored its nutrients and surface water. Still, the absence of Man's etching tools on this plot of earth created a tough skin when he tried to penetrate it.

Abe calculated the approximate number of pounds of feed he would need over the winter to feed the cows, horses, geese and chickens. If nature cooperated, he might harvest six pounds for every pound of seed he planted and feed his family and livestock on a plot of land that was less than a third of the typical American farm. He would need a team of horses, human endurance and patience, and some simple tools, tools that would have been antiquated on a larger farm. With a plan etched in his mind, he hitched his team of horses to his two-furrow riding plow and set about breaking up and turning over two long rows of soil at a time. The scent of buried moisture and the remnants of grazing cattle rose up and mingled with his Turkish tobacco. When the struggle of a sparse winter yielded to mild spring days, the first passes of the plow, for Abe, never failed to sow hope into the earth. This year, he sang aloud to his team of horses with profound joy. When the first field had been transposed by a set of harrowing discs from a weedy meadow into an inviting uniform plot of rich lumpy earth, he mulched it again into a finer, more receptive blanket of earth. Offering a prayer of thanks to God, he then planted his first rows of corn on a plot of land deeded to Abraham Reznik.

  The Reznik family occupied the farm like a foot in a well-worn shoe. In a year, they had sunk enough sweat, worry, imagination and hope into the house, the barns, the paths, the fields, the fruit trees, the grape vines, the cows, the geese and the chickens, that this small corner of the earth became their home and final destination. Hinda stood in the kitchen and took in the swept floors, the scrubbed counters, the half-empty cupboards, the wood-burning stove, and the scarred pine table that witnessed the congregation of her family through the day. She felt something new and strange, yet somehow familiar, embodying the strongest forces of her soul - her obedience to God, her loyalty to family, her humility, and her quiet suffering. She was home again.

Young Symie on the Reznik Farm

Hinda, Si and Abraham 1920

Si with his class at Chambers School 1922
Si, second row, third from the right

The Farm Becomes a Resort

Other Jews would follow. The Reznik home became a tiny magnet, a miniature central depot, for old-world Jews who needed a kosher meal in a kosher house in rural southwest Michigan. Some came to escape the noise, congestion and pollution of the big city with its meanness and depersonalization; others came to explore making the same move that the Rezniks had just made. Whatever the reason, these old world Jews, who laughed together, bantered in Yiddish and tacitly shared ancient memories, found refuge in the Reznik household. The Rezniks opened their doors on several occasions to Mr. Samson and Mr. Liph from the Jewish Agricultural Society while they pursued the placement of other Jewish families in the area. As if a benevolent force had intervened, this sequence of events was creating two possibilities: The Reznik household was becoming part of the hub of activity responsible for bringing new Jewish farming families to the area created the potential for a small Jewish community reminiscent of the shtetl. Also, the lake breezes combined with farm freshness created a lure for observant Chicago Jews who wanted to spend some leisure time in God's country.

At first, the Rezniks accommodated the families of their cousins. They arrived when the days were getting longer and warmer and the trees had leafed new-green foliage; the dormant farm awakened to a commotion of activity for all the Reznik hands. Rather than a burden, their arrival was an honor, a blessing of sorts, after years of having little that anybody else wanted, finally to be in a position where it was possible to extend hospitality. The Rezniks did their daily chores – milking the cows, feeding the chickens, collecting eggs, changing the straw, plowing the fields, shopping for staples and fish – and then simply shared the fruits of their labor with their guests. Their city-dwelling relations enjoyed the fresh warm milk, the freshly laid eggs, the un-chlorinated well water, the solitude, the agrarian lifestyle, unspoiled by the complexities of technology. Nearby they hiked the woods running up to an idyllic clean river. The Rezniks served these early guests without expectations and received in turn heartfelt gratitude and a small gratuity, perhaps five or ten dollars, whatever the visiting family could afford.

In this unpremeditated way the Reznik Resort, among the earliest Jewish resorts in South Haven, formed. This was not the first time that the Jewish Agricultural Society played a role in the formation of a Jewish resort. When the organization helped Jews purchase land in the Catskills of New York in the 1890's, it initiated a series of events that lead to some of the finest resorts in the Northeast. While the land in the Catskills was used exclusively for luxurious resorts, in the country around South Haven, the farm life paralleled and supported the resort life. The Rezniks were a farm family that accepted the role that resorters placed upon them. They did not seek it. The Reznik hospitality spread by word of mouth in the close-knit Chicago Jewish community. With time, as novelty became responsibility, a free visit became a thing of the past. But still the Rezniks offered something unique when the word circulated that $18 purchased a week's worth of home-cooked kosher meals and a clean room in the country.

In time the resort business expanded, creating the need for new buildings and more tillable land.

Joseph Reznik, "the Wizard", the brightest of the Reznik boys, helped his father design and construct the necessary buildings. Of paramount importance was the construction of a dining room large enough to serve the guests all at one time. The dining area in the main house could serve only a few families at one time. The Rezniks looked with renewed interest at the old racehorse barn. It was a large well-constructed building of thick oak timbers that was underutilized, because the dairy barn provided adequate space for the cattle and horses. Together, they visualized a grand dining area with rooms for lodging on a new second floor. Guests could reach the second floor by a flight of stairs ascending outside the modified racehorse barn. They further imagined that if the racehorse barn were to serve as the new dining room, they would need to design and build an attached kitchen where Hinda and her daughters could operate close by. Things were beginning to take shape, at least in their dreams.

Converting a racehorse barn to an inviting, appetizing dining room took place in many stages, but the most difficult one involved eliminating the horse aroma that infiltrated deep into the earth below the floor of the barn. First, they removed all the stalls, changing it from a compartmentalized warehouse to a spacious, but odorous and poorly lit dance hall. Then they pried free the floorboards stained with ground-in-manure and swollen with gallons of urine, saving the good nails and setting aside the wood for kindling or board replacement in the other barn. Then, starting at the back of the building, by hand they dug for untainted earth. Before the job was complete, they had discarded a mountain of acidic, richly odoriferous earth that filled wagon after wagon. The room-sized, waist-deep pit that remained was filled with a shipment of fresh, sanitary sand. Then they covered the past with a new pine floor.

Over time, they tore out the old roof and constructed a loft with eight rooms, topped off with new rafters and a newly rolled asphalt roof. They finished the rooms with new lathe filled with fresh gypsum, skimmed smooth with plaster. To the north of the dining room, connected to it by a short enclosed corridor, the two Reznik men built a new kitchen, equipped with a stove and enough counter space to cook for 110 guests on weekends. Adjacent to the southwest corner of the dining room they constructed a small bedroom where the Reznik parents slept during the resort season (the house was occupied by guests). Three summer cabins, each with one large room and two smaller rooms, capable of lodging more than one family, were erected south of the cinder driveway. Just to the north of the cabins, there were two outhouses, one for each of the sexes, each with three holes to accommodate all sizes of tushes comfortably. Each summer, to the west of the dining room, they set up three large tents. As guests occupied the main house during the resort season, one of the tents sheltered the Reznik boys. Finally, as bottomless pitchers of ice water became a Reznik tradition and food storage a greater requirement, Joe and his younger brother, Meyer, built an ice house just west of the main kitchen.

In 1912, the town of South Haven introduced streetlights; the Reznik farm and the resort, however, functioned without the luxury of electricity and plumbing during their formative years. Kerosene lamps sufficed as they always had. While the farm was churning out feed and produce consistently and the resort was on an upward spiral, the Rezniks saved a large enough sum to make a down payment on a kerosene-fueled Delco generator which charged 24 wet batteries. This system, still a rarity in the countryside outside South Haven at the time, supplied enough juice to wire all the paths and lodgings with the wondrous but stark glow of bare-bulb light. The unveiling of the new system on the Reznik farm became a community event. After dusk's glow vanished over the horizon, spectators from miles around the community admired its effect. In a time when people did not scoff at an Indian Head or a Lincoln copper camouflaged in the dirt, the $600 outlay for the installation of this system was a small fortune, about what a small farm earned in one year. The Rezniks did not waste any pennies obscuring the precious country light with shades.

Later, when funds permitted and fierce resort competition necessitated it, the Rezniks installed running water and plumbing. Joe oversaw the construction of a new room contiguous with the garage. The new construction stood at the end of a two hundred foot cinder driveway that ran perpendicular from the main road between the main house and the orchard before it dead-ended in the garage. The Model T's and the rarer Packards now drove up to the swoosh of flush toilets and the gentle rain of low-pressure showers. Another twentieth century gadget became part of the Reznik landscape – a chugging water pump which slugged down the juice of the Delco generator and treated Chicagoans in the manner to which they were accustomed.

Other improvements on the Reznik farm resort ignored twentieth century invention and relied on timeless human labor and methods that were centuries old. The gathering of gravel from the shores of Lake Michigan at the mouth of Deerlick Creek was one of these time-honored back-breaking tasks. While summer may have seemed the more logical time to haul stones from the shallow, stony bed at the shoreline of Lake Michigan, the team of horses preferred the slick surface of a winter blanket of snow to draw their tonnage. What's more, time, a resource that was scarce in the summer, became more plentiful during the winter. Instead of dialing three numbers and placing his order for six yards of gravel to shore up the pathways and create gravel patios, Abe Reznik, reared on old-world feats, waited for the ideal winter day to harvest the stones. The expedition took place after the first snow deep enough to bury the branches and ruts that impeded the runners of the sleigh, and when the lake was still free of its glacial edge. He waited for the storm's gale with its onslaught of thundering waves to subside. When the coldest day gave way to a still interlude and rebounding temperatures, and the lake became placid, he hitched the horses shod with spikes to a unique four-by-four foot box mounted on the sleigh. He constructed this box of strong timbers placed at even intervals to allow Lake Michigan to seep out from the haul of gravel.

A group of farmers with their special bins aggregated in the broad expanse of shallow, chilly water at the mouth of Deerlick Creek. Prepared with shovels and the hydraulics of straining quadriceps, they planted their hip high boots in the stony bottom. Then, with an awkward wrench to the spine, they hoisted the gravel weighted with water and sand onto the bed of the sleigh. As the pile of stones mounted, the runners of the sleigh sunk into the floor of the creek where it dead-ended into the lake. Had any farmer naively ventured out alone on this expedition, at this point, he would have been disappointed to learn that the gravel turned his sleigh into an immobile planter rooted in the sandy gravel. The same team of horses that skidded the empty sleigh to its destination could not budge its haul. One by one, the farmers broke the tenacious grip of the lake floor by hitching a second team of horses to his sleigh. With the team of horses borrowed from his neighbor, the sleigh was easily drawn to a level field of snow. Here, a single team of horses took over the ponderous sleigh and their decision to harvest the stones while bundled in woolen caps and scarves, instead of bare-chested, paid off.

By the time Abe reached his farm, after traversing the quiet wooded paths near the creek and passing through the bustling trade on Phoenix Road that gradually opened into snow covered pasture land before reaching his farm with the gravel, what water had not glazed the gravel seeped through the bed of the box. The ingenious design of the box permitted an easier shovel-free unloading than the exhausting loading. One of the Reznik men stood at either end of the last timber of the bed of the box and began to rock it, first freeing one end and then the other. The gravel supported by the timber would tumble to the ground. They jockeyed each timber of the bed in the same way until the entire load lay piled on the ground. There it remained until spring revived the activity of another resort and farm season.

Abe savored the ache and fatigue left in his limbs by the days' work, because it connected him with the world gone by that he knew best. What little money he had saved with the hand-retrieved gravel would allay ten-fold the amount of worry and guilt he would have had had he spent the money. There may have been an easier way; he had done it the way that felt right.

FRUIT FARM   AUTO and BUS
HOME COOKING   AT DOCKS

ELECTRIC LIGHTS

Reznik's Resort
SOUTH HAVEN. MICH.

PHONE 30 F 3   R. F. D. NO. 5

Making More Farmland

Reclaiming the lowland of Trowbridge flats from the straggly woods and thickets, which covered it, was another ongoing project demanding perseverance. A hemlock forest common to southwest Michigan, it was believed, had once covered the Trowbridge Flats. Centuries earlier, the neighboring Black River had consistently flooded the land, suffocating these fernlike, august trees. Limbs fell first from the rotting trees, but over the decades, the trunks slowly subsided in the earth until their crowns were swallowed by the muck. Then centuries of repeated floods left a layer of rich fertile silt and converted these tree carcasses into soft coal-like chunks of organic matter laced with gristly areas hard as slate. When Abe Reznik decided to alter the land for his purposes, whether these events occurred or not, the theory seemed to explain the origin of some of the adversities he encountered.

Abe did not attempt to tackle the bogs of Trowbridge Flats until the summer sun lowered the earth's water table, making the land more navigable and workable. The first step in reclaiming the land was to clear it of its natural inhabitants--brambles of blackberries, thickets of the pervasive weedy sumac trees intermingled with stunted white pines, thirsty swamp maples with sprawling superficial roots, and rotted black birches that feasted before dying in the moist soil. Around the brush and the smaller stumps, he wrapped chains and attached them to his sturdiest team of horses, exhorting them to pull until they freed the roots from the loose, wet silt. For the larger stumps, he borrowed the amateur expertise of a neighbor who knew his way around dynamite. When the surface impediments were removed, a pitted, uneven, boggy wasteland remained.

Next, he set about changing the stripped flood plain into arable land. With his deepest plow blade, he pried air into the depths of the soil and brought the lower wet layer to the surface where the sun could bake it dry. The plow blade rode deeply in the earth with minor resistance until it suddenly snagged like a fishhook on a corpse. With such ferocity, the team of horses was jarred to a halt that their front legs were lifted into the air. When they came down, they began thrashing with all four legs to free themselves. At times, rather than budging the plow from its snag and moving it forward, their hooves, like frenetic tiny shovels, dug them up to their waists in the boggy earth. Legless, their long necks tossing from side to side to lift themselves from the mud, only driving themselves still deeper, while snorting and neighing helplessly, Abe unhitched his team from their tow and headed for Bill Trowbridge's Percheron horses. He bypassed his other team of horses because he knew they would not work with their stable mates to their rear, as if these terms were in their contract. Besides, Bill Trowbridge's heavy draft horses never failed to effortlessly pull their snorting weaker neighbors from their mud trap. When the thrashing animals caked like pigs were freed, Abe left the wallow they created uncovered so that it might dry in the sun and so it would not ensnare his beasts again.

The work was tedious and difficult, taking place over many seasons, but it proved fruitful. The fifteen new acres of arable land nearly doubled the tillable land. What's more the soil was unusually fertile and the pampered crops doubled their normal height. Millet topped out at six feet. It grew tall enough that the plants of one row gradually leaned to join tassels with neighboring ones, forming tents and tunnels in which the children played hide and seek and Robin Hood. More acres of feed crops allowed a larger number of cows to live robustly over the long Michigan winter. Gradually, after years of turning the soil over spring and fall, enough moisture saw the light of day to create rich safe fields.

Life on the Farm-Resort

While the farm and its surrounding land with its natural resources, timber, pasture land, and rivers allowed the Rezniks (in large part) to be self-sufficient, it did not make them prosperous. By the end of an average growing and resort season, the farm-resort generated only five hundred dollars. This amount covered the expenditures of the non-productive period between Labor Day and Memorial Day and the expenditures of any planned improvements and routine maintenance. Each year, for beef, they slaughtered a calf and most years, a cow that had outgrown its usefulness. They raised geese and chickens for meat and eggs, and fished in the nearby river and bayous. In addition to the feed crops for the livestock, they grew some fruits and vegetables including potatoes, onions, grapes, pears, apples and peaches. The livestock gave them milk, which they turned into cheese and butter year round. They acquired fuel for heat and cooking from the timber of unfarmed acreage across the road, refrigeration and ice from the river, gravel from the lake, wine from the vineyard, lumber from the sawmill on the adjacent Newton property, feather pillows from the goose down, and ointments from the goose fat.

They had other needs however, for which their land did not provide. Shoes tattered from an on-foot existence had to be repaired, and rarely replaced. If a pair of overalls or an overcoat could not be darned and refurbished, it had to be replaced. Coffee, carefully doled out each day, was a necessity during the winter. They required refined flour to turn the homemade coarsely ground grain into edible loaves. Regularly, Hinda Reznik shopped in town to acquire these staples of flour, coffee, sugar and fish, but only rarely was she enticed by a treat.   It was an existence that rarely diverged from the necessities.

And even then, the five hundred dollars was not always enough.   Some time after his father died when he had returned to South Haven after testing the waters of Chicago, Si received word unexpectedly that Fred Radke at the bank wanted to see him.   Since he had no recent business at the bank, he was bewildered about what the bank wanted from him.   He later learned that among some old papers in storage, Fred had discovered an old banknote for fifty dollars, recording a loan his father had taken.   Abe had long since passed on and Fred was sharing his discovery, not collecting on it. Si suspected that the banknote attested to an off-season when the five hundred dollars did not stretch long enough.

During the winter season, the Rezniks returned to the main house that they eventually referred to as the "winter house". The only source of heat during the frigid and blustery days of winter was a wood-burning stove downstairs near Abe's room. Upstairs, the only furnaces were humans, insulated by layers of horse blankets beneath a homemade down comforter.   All three Reznik boys shared their heat in one bedroom. When temperatures outdoors reached fifteen degrees below zero, the trek to the outhouse in the wee hours of the morning was dangerous, especially for a young child. Symie, scrawny as a young child, rather than face the unbearable early morning chill, regularly opened the second story window and, with his brothers' permission, relieved himself. After three months of this assault, the urine eroded a hole in the screen large enough that it required replacement. Downstairs, after arising, Abe stoked a new fire in the stove and warmed himself for the next hour while he cleared his lungs of the overnight accumulation produced by his two-pack-a-day habit with Turkish tobacco.

Sustenance

While conditions were at times harsh and at times unpredictable in the Rezniks' daily and seasonal life, they always had plenty of food on the table. They ate a "milchig", or dairy diet, the natural outcome of Abe's orthodox Jewish beliefs and of living on a small dairy and poultry farm. The majority of their meals consisted of homemade cheese, milk, eggs, bread, butter, and produce such as potatoes and greens, accompanied by a variety of fish including homemade gefilte fish and trout from the Black River, if Si and his sister had a good day with their fishing poles. On special occasions, such as the Sabbath, they would eat a "fleishig" meal, a meal with beef or poultry without the usual dairy foods. For these meals, that drew from a store of slaughtered geese, the meat of an old cow or the annually schected (ritually slaughtered) calf. Many nights, Abe cleansed his palate with a glass of homemade wine. Mostly their nourishment was homegrown or homemade, coming from their fields, their barn or their "blue room", a cool room where Hinda Reznik made her cheese.

By the time Abe had reclaimed the lowlands, he was farming approximately thirty acres. He devoted most of the land to feed crops for the animals, hay, oat straw, millet, timothy, corn and fodder. A minority of the land, he devoted to foods, which might embellish the plate of a resorter or break the doldrums of fish, cheese, milk and potatoes on the Reznik fall and winter menu. They grew strawberries, blackberries, currents, red and black raspberries, cantaloupe, and watermelon. They had the vineyard harvest as well. They also grew staples of scallions, green vegetables, winter onions and potatoes, the early potatoes for the resort and the later ones providing starch for winter. The potatoes and greens, they stored either in an outdoor winter pit for use in the spring or in the Michigan basement of the winter house for over winter use. Finally, they tapped the abundant maple trees and boiled the sweet sap for syrup.

The fattening of thirty geese during the fall was not only a yearly tradition on the Reznik homestead, but a matter of winter survival. From the flock of geese, they drove thirty suitable ones into a coop in the barn for fattening where they could share the body heat of the horses and cows. A few geese with a gander they set aside to propagate the next winter's rations. The domesticated geese accepted their fate quietly, having lost through generations of breeding their urge to migrate and announce their presence and departure with their familiar honk. They fed on cracked corn ground at the Bangor Mill, and stored it as meat and fat. In their tight quarters, they wasted very little nourishment on movement. Hinda periodically stroked beneath their feathers to test for plumpness. When she found a thick enough layer of muscle and fat, the geese were freed from their pen for a jaunt to the nearby Black River. At this stage, all the Rezniks participated in driving them through virgin snow to the river in a sort of purification ceremony. Each of the geese had time to baptize itself in the river before the Reznik boys stationed on the other side of the river drove them out of the water toward their old quarters. Their soiled pen, meanwhile, was cleaned and a couple of feet of fresh straw were placed on the floor to filter any new excretions and to preserve the cleanliness of their feathers.

During their early experience on the farm, the closest schochet, a kosher butcher who was often a

Rabbi, capable of slaughtering the geese for kosher consumption lived in Benton Harbor. Abe and the Reznik boys arose at four o'clock in the morning to begin the twenty-eight mile journey to the Rabbi. To preserve its cleanliness, they hand-carried each goose to the back of the sleigh. In Benton Harbor, the schochet, with a single draw of his perfectly honed challef, or knife, slaughtered each goose and let its blood drain completely from the severed jugular. By the time he completed his religious task, the sun had long since set. The Rezniks completed their journey as it had begun, alone on a starlit or moonlit road with the arboreal shadows thrown across the snow-packed path. By the time they arrived back in South Haven, the geese, drained of their warming blood, were partially frozen into all imaginable grotesque poses.

All night, the Reznik women plucked the feathers that had anchored themselves into the gooseflesh on the way home. Kosher dietary laws precluded them from dipping the birds in boiling water to relax the gooseflesh's grip on the feathers, as heat might congeal the remaining blood. The Reznik women first detached the wings of the birds keeping them intact. Each wing would become a fresh feather duster for a summer resort room. Then they harvested successively easier layers of the feather coat until they came to the easiest, the down. Hinda Reznik then cut the geese into pieces. She koshered the geese pieces by first washing them, then soaking them in a large tub filled with salt water for a half hour, and finally setting them on grooved board slanted to drain the diluted remnants of blood. After she thoroughly salted all sides of the meat with coarse kosher salt, she placed the large pieces, the legs and the breasts, into one barrel and the smaller pieces, the necks and the wings, into another. Abe then partially buried the barrels outside to take advantage of the competing forces, the freezing of the winter air and the thawing of the unfrozen ground. Every Thursday or Friday, Ms. Reznik would use an axe to separate a portion of goose parts and bake or fry them for a fleishig, or meat containing, meal. Not only did the geese supply a season's worth of meat for the Reznik family and feathers for pillows and comforters, but for Ms. Reznik, an ardent believer in the healing powers of goose grease for any ailment, it provided an unlimited supply of salve for the unending chain of nicks and bruises sustained in farm life.

Before a kosher butcher shop opened in 1922 in downtown South Haven, the Rezniks and other Jewish families that strictly adhered to kosher dietary laws could not depend on having a supply of kosher beef for the winter. Most years, there was an old cow for butchering that either had outgrown its usefulness to breed and give milk, or suffered excessively from arthritis or other ailment. When such a cow existed, it was brought over to the Rezniks where Rabbi Schwartz would perform the ritualistic schecting. After he ran his pinkie along both sides and then down the center of his challef to detect any minute snag, with the assistance of one of the older Reznik boys, he schected the venerable beast with one swift stroke. After he drained the animal's blood, the Rabbi split open the chest cavity and reached into its recesses. He ran his hand between the outer cover of the lung and the inner cover of the chest wall in an attempt to detect any threadlike attachments called cirches that might exist between these two layers of tissue. If he found an attachment, he would not remove the lungs.

When Rabbi Schwartz found an attachment, he had to remove the lungs carefully without tearing them.

He made certain that these threadlike adhesions were separated near the chest wall attachment, as far as possible from the surface of the lung. After the Rabbi had severed all the attachments, the fastest Reznik son raced into the kitchen with the lungs wrapped in a discarded blanket to preserve their heat and placed them on the kitchen table. A bucket of warm water was set beside the pair of lungs that had been carefully unwrapped. The eldest Reznik son moved to the head of the table and put the animal's trachea to his lips and inflated the still warm lungs. Simultaneously, the Rabbi splashed water at the sites of all the thready attachments. If the Rabbi noticed any air bubbles, their work had been for naught as the beast was designated "trayf", or torn, and was set aside for gentile consumption.

If he found no air bubbles, the Rabbi began the tedious process of removing the cirches. The Rabbi, who made his living as an optician, wore a special pair of magnifying lens as he gently peeled away the defects with his fingernail. After each layer of the cirches was removed, the Reznik son re-inflated the lungs while the Rabbi lapped warm water over the site, checking for minute punctures. Only when each circh was removed completely without a leak was the process complete; only then did the Rezniks have beef for the winter. Rabbi Schwartz saved most of these old animals for kosher consumption. As was the case with other kosher dietary laws, the practice of identifying and removing cirches may have had a secondary medical benefit – it was commonly believed that the cirche was the first sign of TB.

After the cow was saved for kosher, Sanderovich, a shoemaker by trade, used his leather-trimming tools to butcher the cow. Any place where blood coagulated, the meat was discarded. Eight to ten Jewish families, including the Rezniks, the Weisbergs and the Schwartzs, that strictly adhered to kosher dietary laws divided the different parts of the animal. Each family received an equal cut of the beef and a piece of the windpipe, spleen, stomach lining (tripe), utter, tongue and liver. No part of the animal was wasted. To this day, Si has fond memories of chewing on a piece of the windpipe used to flavor the juices of a pot roast.

Kosher beef was the entrée on the Sabbath and other special occasions. But cheese, over the years, provided a more durable and readily available form of sustenance on the Reznik farm-resort. Hinda oversaw the production of cheese in a room off the kitchen referred to as the "blue room", an unheated room in the winter house that provided lodging during the summer for the visiting third generation Rezniks. The room had acquired its blue hue and its name from its regular coats of calcimine used to hide the summer scribbles of the Reznik grandchildren. A warm room in the summer, in the winter it settled at the ideal coolness to permit a wooden pail full of soured milk to complete its transition to a hard, crumbly hunk of farm cheese.

Soured milk without the benefit of a curdling agent such as rennet sat in the unheated room until it curdled. After mixing in a few pinches of coarse kosher salt, Hinda poured the embryonic cheese into a large porous bag, an old sugar or flour bag. She suspended the bag over the pail that caught the whey, beginning as a brisk trickle and eventually tapering to an imperceptible ooze. When the ooze stopped, she placed the hunks of nascent cheese on a table covered with oil-cloth and while still in the filter cloth, she placed a dense heavy stone on top of the cheese. The weight coerced more liquid from the solid curd. Hinda then removed the cheese from its filter cloth, condensed but still moist, and placed it on the table to dry and age. For two weeks, she exposed the different sides of two to four pound hunks of cheese to the cold dry winter air modified by the walls of the blue room. Each time she entered the room, the rancid ripe-feet aroma clung to her nostrils until she cleansed it with the pristine arctic air outdoors. Having taken three to four weeks from curd to a beige table cheese, the final product had a shelf life of many weeks.

With an ice pick, several times daily, Hinda chiseled small chunks of cheese from the brittle block that challenged the blade of a knife. Cheese was the backbone of many meals. It was produced from the most readily available product on a farm, transformed into a product that lasted longer than it's parent, more sensational to the taste buds and more robust in nutrients. The Reznik family took advantage of this natural transformation of milk in the same way farm families with domesticated animals had for four thousand years.

If Hinda was the family expert on cheese making, Abe was the family authority on winemaking. The small vineyard that they inherited had red, white, but mainly blue grapes, all of which matured at different times during the summer. Some of the grapes Hinda simmered into jam while others refreshed the summer resorters and the Rezniks themselves. Symie passed through the vineyard on his way to the lowlands to herd the grazing cattle at the end of each day. When in season, he grabbed bunches of grapes and devoured them as he chased after the lumbering herd and drove them home. Most of the grapes matured in the fall after the resorters had returned to their fettered lives in the city. These grapes succumbed to a family grape-picking

69

expedition that occurred in September before the first frost. Their harvest usually filled two fifty- gallon barrels.

From this point, Abe took over the fate of these grapes in his spare time. Winemaking was a labor of love, suffused with guilt. Periodically, using a wood-handled stomper, he pulverized the grapes in the barrel, shredding the skins and pulp, freeing the juice. Over a month or two, into the Indian summer days of autumn, the juices fermented, disturbed only by Abe's stomper. Just before the first prolonged freeze of winter, he added the sugar to the mash. Then, as in cheese making, a large heavy stone helped separate solid from liquid, only in a reverse fashion. In winemaking, a broad heavy stone compressed the pulp to the bottom of the barrel so that the fermented juices could be drained from a spigot placed a hand length from the bottom of the barrel. From two fifty-gallon barrels of grapes, Abe drained half as much wine that lasted until the next winemaking season.

One year, while three barrels of grapes sat swapping some of their sweetness for spirits, Ray Symonds, a Delco Light salesman, discovered the homemade production. He was surveying the grounds and sketching plans for a future Delco Light system when he found the three barrels behind the barn. Symonds, who had a reputation as a shikker, a drunkard, inquired about the future of the barrels of grapes and maneuvered to claim a barrel for himself. He learned that only the sugar was holding up this batch of wine. "How many sacks of sugar do you need?" he inquired. Sugar was the most expensive ingredient in the wine, not found on the farm or in Michigan, so Abe expected it was the idle chatter of a shikker. The next morning, several paper-lined burlap sacks filled with sugar arrived at the Reznik farm with a note attached which read, "Not too much sugar in mine".

Another autumn, the Rezniks learned a lesson about the pulp that settled at the bottom of the barrel. It seemed a shame to waste the chaff of the winemaking process when there were so many creatures on the farm that might enjoy fattening themselves on it. They spread the grapes' skins in the chicken coop, and each of the chickens gorged its gullet. Gradually, they became a flock of winos, cockeyed and stumbling around the pen. Many lost their balance, tumbling awkwardly on their backs and sides in poses unfamiliar to fowl and reminiscent of the headless frozen geese that became grotesque sculptures on the way home from Benton Harbor. The next morning, the chickens, upright again, were clucking about a nasty hangover. After that fall, Abe disposed of the grape skins where only a shikker raccoon might indulge itself.

The fifty gallons of wine, drained into bottles reused from year to year, lasted four seasons, taking part in a variety of holidays, occasions and habits. A bottle or two were drained filling the four glasses of wine during the Pesach Seder. An occasional bottle changed hands when they were invited into someone else's home. Most of the wine, however, served what became an evening ritual for Abe. Before the evening meal and after a parching day in the fields, when his mouth was plastered with dirt and stained with cigarette tar, Abe needed to cleanse his palate. He took the open bottle of wine to the room beside the kitchen and filled his embattled oral cavity, savoring its homemade flavor in every cranny. When the stale tobacco and the toil of the day were fleetingly a memory and the taste of the wine momentarily at center of his universe, he swallowed it in two gulps. He closed the bottle, came to the kitchen, and said the prayers for the evening meal.

Two Icy Jobs

As the resort business grew and the farm rooted itself, the Rezniks required more refrigeration space. Joe and his brother, Meyer, met this need by building a simple structure called an icehouse, a sort of walk-in refrigerator capable of holding an almost limitless supply of ice. It was two stories high, but with a single floor the size of a large living room. For insulation, they lined the entire structure, including the doors, with three-inch thick sheets of cork. On the east side of the building facing the summer kitchen door, they constructed a large main entry door and on the west side, a diminutive door, only four-foot square, halfway up the structure. When the ice was piled high enough on the west side of the icehouse that one could no longer approach the ice stack from the main entry, the smaller portal of entry was used. Each year Abe Reznik accumulated a large mound of dirt beneath this door to create a ramp for the sleigh. When burdened with ice, Abe could take the sleigh right beside this door and unload it with greater ease.

The mining of ice each winter took place before the spring thaw when the ice reached its greatest thickness. The job was dangerous and not for the fainthearted. Years earlier, Elmer Lammssies was hauled frozen stiff and lifeless from the Spring Hill Pond after he lost his balance lifting a cake of ice from the winter water. Abe believed the task perilous enough that he would not jeopardize the more plentiful years of his sons. The elder Reznik, therefore, did the job alone. Clad in thick canvas overalls, he rode in his sleigh pulled by spike-shod horses to one of two places, the Black River at Trowbridge Flats or to a spring-fed pond two miles west, the spot Lammssies slipped beneath the ice. (The same pond supplied water for the bottling factory in town during the summer.) After surveying the expanse of the ice and testing its strength, he parked the sleigh and horses on to the river or pond beside the area he planned to cut. If he was mining virgin ice, he drilled a hole through the ice with a large hand-held auger wide enough to fit a large ripsaw with ferocious teeth. He gauged the thickness of the ice and estimated the block-size he could manipulate out of the water onto the back of the sleigh. On average, each block was seventy-five pounds and between a foot and a foot and one-half square. After cutting two sides of the first block, he augured another hole so that he could cut the remaining two sides. Using a pair of crude, but heavy-duty pair of ice tongs, he teased the first block from the mile long sheet and put it in the bed of the sleigh and shoved it forward. The most difficult block behind him, he lingered over the small dark square in the vast sheet of ice; then he was back to work; many tons remained to go for the season.

If Abe was lucky, another intrepid cutter had already broached the ice field. Then, if there had been no intervening thaw, he could park himself and begin to cut where the other left off with certainty that he would not crash through the bed of ice. Yet he was always wary of the ominous crackling sound of splitting ice. He considered each step, as if his family depended on it. He could not afford to allow a foot to slip out from under him or to lose his balance and pitch into the breath-taking cold of the river. When he had made it safely to the end of another day, the sun dropping below the trees and the back of the sleigh filled with cakes of ice, he would hesitate before returning home and permit himself a moment of intense gratification.

He scanned the long stretch of terrain that sloped up from the frozen river bank-- conifers that struggled to keep their winter green in the dying light, endlessly dividing branches of poplars, maples, beeches and oaks silhouetted against dimming skies, a quilt of snow that knitted the river with the terrain it carved, and united it with the icy blue skies beginning to deepen on one horizon and pale on the other. The saw silent, sporadic sounds amplified his solitude, the crackling and shivering trees, the river lapping against the man-made shelf of ice, the occasional snort of impatient horses and, when all else rested, his own wheeze. And he thought how the ice was free and the way in which God had allowed him to be self-reliant, because he thrived on hard work; he knew nothing else. Instead of waiting for the world of GE, he embraced the practices of his old-world ancestors.

Back at the icehouse, Abe unloaded the blocks of ice so that there would be no danger of an avalanche and the blocks would be easily accessible when it came time to use them during the resort season. He stacked the blocks into tiers, making an uneven set of steps that he could climb, the lowest step nearest the main entry door and the highest step on the far west wall near the smaller door. He spread sawdust over each layer of ice, not only to provide traction, but also to maintain the distinctiveness of each block and to prevent slipping between blocks. Enough room remained free of ice so that Hinda could store perishable food and drink.

Handling the large blocks of ice in the icehouse was hazardous enough that it too was left to the eldest Reznik. As he could reach the back tiers of ice only by climbing the treacherous lower ones, he regularly faced the possibility of taking a limb-breaking tumble, especially in the summer months when the room temperature rose a little and the surfaces slickened. Periodically, Abe had to shift around the blocks of ice so the Reznik women, who came and went frequently from the icehouse, could access them safely. Among their responsibilities, they replenished the smaller insulated oak refrigerators located in the dining room and the kitchen with blocks of ice.

Every summer, resorters quenched their thirst with un-chlorinated well water chilled by nuggets from the frozen crust of the Black River or Spring Hill Pond. Each day, Abe freed a block of ice from the ice pyramid in the ice- house. With an ice pick, the Reznik women chiseled fist-sized chunks from the blocks and stacked them in a fleet of iron pitchers. They filled each pitcher with the well water, gathered them on trays, and then delivered the daily allotment of ice water to each room around the resort. When the lake breezes took a siesta, the cicadas purred, and perspiration glossed every limb, the ice water became a simple country treat.

One chore for Hinda that was more onerous during the winter than during the summer was washing clothes. During the summer, when the load of clothes was armfuls greater, the winds were gentle and mild and the water into which she constantly dunked the clothes, while her brow beaded with sweat, delivered a welcome coolness. During the winter, conditions were more suitable for food preservation than for washing clothes. She waited for the sun to reach its highest point in the sky and then, in the lee of the wind, put a large galvanized tub and washboard in its weak winter warmth. By the time she had rubbed and dunked and rubbed and dunked each article of clothing, a crust of ice rimmed the tub of water. After washing each piece of clothing, she hung it on a clothesline where it would slowly solidify before it dried. The limp pile of clothes that she took out to the washtub, she returned to the warmth of the winter house, two or three unwieldy stiffened shapes at a time. When she completed her task, as if the assault of the freezing water were not enough, she collected a handful of snow and rubbed it into her raw red hands and arms to restore their circulation. Anyone who saw her at that moment knew she had just completed a week of winter laundry. **

Winter was the ideal time to harvest the dead standing lumber from Trowbridge's woods across the road or from Newton's back forty acres. The needs of the farm hibernated and the foliage that lay in early decay on the forest floor instead of rustling in the wind did not impede the fall of a dead tree. One winter, the Reznik men accumulated a bounty of wood, one hundred cords, chopped and split, all ready to use in the kitchen for heat and cooking. They left the wood temporarily in the snow near the banks of the frozen Black River that ran through the back forty acres. That year, cold northerly fronts kicked up regular snow

---

* *Hinda's belief in the restorative powers of fresh snow extended beyond her own ailing limbs. One particularly frigid day, Meyer had to retrieve the milk cans from the barn. On the way from the barn to the house, one of the cans tipped over when the sled used to transport them hit an uneven patch of snow. Meyer, encumbered by thick winter layers of clothing, had to struggle to free the can filled with milk from the snow and to right it so that it would not happen again. By the time he reached the warmth of the house, he had spent too much time in the winter gale. Hinda noticed right away that one half of his nose was glowing red while the other side had blanched white. Without wasting a moment, she ran outside for a handful of snow and slapped it on Meyer's nose and rubbed it until the circulation returned.

squalls off of Lake Michigan, leaving a mantle of snow that would challenge any spring thaw. Eventually, however, the spring thaw took its toll and turned the deep snow to deep puddles and to a rising Black River.

The Rezniks, meanwhile, turned their attention to the possibilities of spring and their own fields, counting on the lumber stored out of view of their usual daily rounds. One time when the Black River at the Trowbridge Flats spilled three feet above its banks, they checked on the long wood stack and found that it remained at a safe distance from the river's grasp on the high land of the back forty acres. Nevertheless, after a day of heavy spring showers left small ponds and streams in the unpaved road in front of their farm, one of the Reznik boys decided to check the Black River on the back forty. He ran home to deliver the news that the river had swelled and caught their winter's work in its current. Emerging from the woods, he yelled, "It's taking it. It's all going." By the time the rest of the family arrived at the banks of the river, there was nothing left to save. Unhappily, they took a ride into town along River Street and watched their 100 cords of wood among all sorts of human items and debris float downstream, seized by the acquisitive muddy river.

Medical Care Penoyer-Style

Old Doc Penoyer delivered medical care at the right price for the struggling farmers—all farm fowl were accepted in trade from those who could not meet the one- dollar fee for an office visit or the two-dollar fee for a country visit. As it always has, much of the infectious illness that required a doctor's attention came along in the winter. Winter snow and frigid temperatures made it difficult for the ailing patient to make an office visit. Therefore, the old Doctor Penoyer frequently had to prepare for the worst winter weather conditions to see his patient. He donned his horsehide mittens that stretched above his elbows and draped three blankets over his knees and over his horses. When he finally made it to his destination, before even glancing at the patient, he sidled his backside and his hands to the stove and warmed up, taking his stethoscope out to remove the shock of winter. He usually accepted an offered glass of grape wine and engaged in conversation for a quarter of an hour before he turned to the sick child and said, "What have we got here today?"

One winter day, the wind-whipped snow measured in yards; the mail never left the station, and old Dr. Penoyer had to get to Grandmother Starks. He hitched two spiked horses to his one-horse sleigh and set out into the empty, howling white world with just enough skin exposed to see. Over the party-lined, battery-operated phones, a group of people learned of the journey before Dr. Penoyer. In anticipation of his route, they lopped off the tops of the deepest drifts in the road with their shovels so that he would not get stuck as he passed through. Other curious snowbound farmers who saw what was happening insulated themselves in felt boots, wool caps and scarves, and several layers of trousers, and helped clear the way. In what turned out to be a community-wide effort, Dr. Penoyer made it in time to save the life of his patient. So the legend goes.

After the Great War, old Dr. Penoyer's son, Cecil, eventually joined his practice. When a patient made an office visit, the old doctor would poke his head out of the door leading to the patient-examining rooms and inquire, "Are you here to see the 'old' Dr. Penoyer or the 'young' Dr. Penoyer?" Nothing brought a more heartfelt smile to his face than the answer, "The young Dr. Penoyer." The misfortune, however, was that the years of service that the old Dr. Penoyer had given to the community produced a trust that was hard to overcome. Too often, the answer came back, "The old Dr. Penoyer," and the smile would not open across his face.

Getting Around

Though the automobile revolution was sweeping the country, the most reliable transportation on the farm-resort before the Great War was the horse-drawn buggy in summer and the sleigh in winter. The Rezniks did not purchase their first car until after the Great War. They could not even afford "everyman's car", Ford's Model T. In many ways a car did not have the versatility and the backbone of a team of horses pulling a wagon, buggy, or sleigh. The deeply rutted unpaved roads, repaired without benefit of machine, played havoc on the tires whereas hooves took them in stride. A sleigh drawn by a team of horses slid over unplowed roads to places no motorized vehicle could hope to approach. While a blanket could keep the engine warm on a horse on frigid winter days, the coolant in the radiator of a car had to be completely drained if it was going to be left outside for any duration awaiting the return of its passengers. During the summer, the same coolant had the unfortunate habit of boiling away on hot summer days or on especially hilly or rutted journeys. What's more, Abe, who had from boyhood valued a horse as much as one of his own limbs, because it was at least that valuable, refused to sit behind the wheel of a Model T. So when the Rezniks finally did purchase a Model T after the war for a savings-consuming $600, it was at the behest of the older Reznik boys who could no longer overlook the march of progress. Like the purchase of their land, it was made in an atmosphere of trust and equanimity with Chris Niffenegger. From one honest man to another, he uttered the soothing words, "Bring me some money when you can."

The Model T was a simple and sturdy machine, but it was a "man's" machine that could not be approached daintily; it could snap your wrist just getting it started. Though the electric ignition was available on many makes of cars, the Reznik's first car had a crank ignition that required the strength of the older brothers to start it. Demountable wheels, also available on many models, also fell outside the Reznik budget, making it a more onerous chore to change the frequent, anticipated blowouts on the roads of the late teens and twenties. The auto had a relatively complicated set of foot pedals, the left foot controlling the brake on the far left and the right foot controlling the reverse pedal in the middle and the clutch pedal on the right. The steering column projected at an angle from the floor through the legs into the lap. Located on the steering column, the right hand manipulated the throttle of the car and the left hand, the spark lever (oil drip). The Rezniks could only afford an open model that allowed the grit of the road to coat their teeth and clog their nostrils, the unexpected spring showers to drench their overalls, and the glaring summer sun to broil their backs. Yet the Model T brought progress, placing them in the mainstream of American life and making a city like Lansing, a distance of one hundred fifty miles, eight hours and several blow-outs away, instead of an insurmountable buggy ride or a plodding ride on a milk train.

The Model T was relegated to a farm machine after the family purchased their second automobile, a Chevrolet, a step forward in prestige and convenience. It was a more humane machine with a roof that protected passengers from the vagaries of weather, merciful demountable wheels, an electric transmission, and a less complicated and noisy gearshift -- a vehicle that both sexes could manage. Three miles down the road, on the edge of town, they bought the car at the Chevrolet dealership. The model T was used around the farm like its primitive cousin, the horse-drawn wagon, to haul seed from the lowlands to the upper fields. The imitation leather seats, once tended with the greatest care, became faded and worn, coated with field dust and splattered with mud. Yet it was just as natural to use the Model T this way as it was to tour and shop in the slightly more august Chevrolet. Abe may have acknowledged the progress his family made in status and Americanization with the purchase of these vehicles, but his place remained until the day of his death at the reins of a team of horses.

Two Evening Routines

Abe had developed the habit of smoking cigarettes filled with powerful Turkish tobacco as a young man, using the dying ember of one to light the next. At first, he was trying to stay warm in the Russian military, but the habit never left him. Each night on the Reznik farm, without fail, often in twilight wakefulness, he would caringly assemble fifty cigarettes to satisfy his habit for the next day. An importer named Pritkin in Chicago sent the Turkish tobacco to the farm in fifty-pound crates. Roughhewn boards protected the precious aromatic leaves in their travel across the lake. When Abe pried off the outside boards, a room-clearing acrid aroma enveloped the room. Even in the adjacent room, it stung the children's eyes and forced them to seek fresh air two rooms away.

Within the crate, the tobacco was packed in one to two pound pouches of lead foil. Abe carefully unwrapped one of these pouches and placed its contents on some newspaper. Though the tobacco leaves were shredded, the texture was too coarse to fill the narrow cylinder of his empty cigarette papers. He, therefore, kneaded the shredded leaves between his fingers until they were malleable enough to stuff the narrow tubes with the proper density of tobacco. The process of preparing his tobacco and making his cigarettes was one, by necessity, done in solitude, too offensive to all the unstained airways in the house. When he was not smoking in earth's open arena, many of the cigarettes, he lit and smoked in an empty room.

During the evening, Abe also took a moment to rendezvous with one of the few possessions he brought with him from London to the United States and kept until the day of his death. Besides a few tools that he had made himself, among them a level and a wooden plane, he transported in the most protected part of his trunk an English chiming clock. In the winter house, each night, as predictable as nightfall itself, he wound the clock and made sure time would be measured the next day. The grind of the turning crank carried throughout the house. Muted by its journey up the stairs and through the ajar door into the darkness of his room, it fell on Symie's youthful ears that were awaiting this signal to slumber like an alarm for sleep. This sound, man-made and purposeful, was a more important marker of time than the brief ring the clock emitted five minutes before every hour.

Symie's World

When Si, or Symie, his childhood name, arrived in South Haven with his family, the figure-eight roller coaster looming on the other side of the Black River was a nonsensical, noisy contraption that offered no particular fascination to his one-year-old mind. His mother and his mothering sisters were the center of his world. He never did experience the thrill of the coaster making his stomach surge to his mouth, but he did grow to appreciate the quiet, at times distant man, who bore the mystery of a past to his developing eye. He grew to appreciate the hard-working honest man who seemed to be the glue not only in his family but also in the slowly expanding rural Jewish community. As he went from child to man, his mother continued to be a source of security and quiet strength like an oak rocker, but the image of his father approached heroic proportions. For a boy who valued his pennies, Symie was rich in the key ingredients required for a healthy passage through the gauntlet of childhood. Within the recollection of his childhood years, other figures loomed larger than his own and were the protagonists in the story of his childhood, but he does remember some of the deeds and events of his years on the farm-resort.

When the resort was in its formative years and Abe was transforming the farm from abandoned to productive, Symie was a scrawny lad full of good intentions, but with little brawn to put them to use. However, his pint-size did come in handy when someone had to squeeze into a cramped space that most children his age had outgrown. One tight space existed on his neighbor's farm beneath the perches of the egg-laying hens. For three shiny pennies every two weeks, Symie contorted himself to fit in the tiny space beneath the chickens and evacuated the piles of guano that accumulated since he last had been there. Besides this small nasty job, which Symie willingly performed, Symie earned a few more pennies in tips around the resort. As the Reznik resort was located three miles from the beaches, too far in many cases for a day trip, many of the guests swam instead in the Black River. On sultry summer days, Symie gladly guided the guests along the path to the best and nearest swimming hole. As he pranced along the path, he chattered as if all he saw was his to share, the fishing holes, the grapes drooping on the vine, or the bike he saw advertised in the paper. Other times, if something needed fetching, food, water, or something in town, Symie, the least indispensable, was the perfect errand boy. His willing nature earned him a few more pennies.

Symie tried to make contributions to the daily farm work, but his efforts were frequently regarded as an obstruction rather than an aid, particularly by his older brothers. When the United States entered the Great War, his oldest brother, Joe, met the height requirement for enlistment by only one-quarter of an inch. Yet, his frame was solid and muscular from the strenuous chores of farm work and maintaining the resort. For the child, Symie, he represented a compassionate, but commanding presence, someone who had to be reckoned with, but also someone with a sense of humor who could be teased, nagged and cajoled. On one of those mild rainy days on the farm when no other chore was possible, Joe and his brother, Meyer, set out to mend fences. Symie tagged along, hoping to make himself useful. He harangued his brothers in his shrill voice, demanding a promotion from errand boy and guide, to apprentice carpenter.

Whenever he nudged and implored his brothers, he was told to get lost and was addressed in his brother's endearing term, "Get lost you Little Shit." As many times as he tried to wheedle his way into the fence-mending brigade, he had to endure this term of affection. Finally, Symie decided to put his kavetching energy to better use. He hid behind a lilac bush and composed a song for his tyrannical brothers. To the tune of Sweet Adeline, he sprung from the rear of the lilac bush and, bristling with pride, sung to his brothers,

"Sweet Adeline, Meyer and Joe kush mere in tuchus arine."

Loosely translated, his rhyming verse told his brothers to put their lips to the rear of his britches.

One day, Symie's attention lit on a bicycle advertised in the paper. Purchasing the bike grew to an obsession even though it was a luxury his family could not afford. What could he do with a bike on the farm and the surrounding rutted roads? But his childhood imagination that easily disregarded important points of reality compelled him to ask his older brother, Joe, if he could buy the bike. Joe, a practical, intelligent man who raced through high school in three years, said, "No. We can't spend five dollars on a bicycle." But Symie would not be denied; it would take more than a torrent of "no's" to douse his desire for the bike. Finally, under Symie's incessant barrages of imploring and promising, Joe looked around practicality into the heart of the little five-year-old boy that had a yearning. Symie had already promised to pay for it out of the savings he squirreled away in his piggy bank from tips and cleaning guano. The change added up to just enough. While the money was in the bank because he had earned it, Si, in his advanced years, still perceived his brother's gesture of looking into his heart and giving him his permission to buy the bike, as a momentous gift.

With the exception of one bike ride home from a friend's house, Si recalled little about the bike after he bought it. He was pedaling east down Phoenix toward their farm into a veil of whitish-gray. Chasing him from the rear was a throng of scurrying black clouds that flashed at their interfaces and began to scatter grape-size drops on thirsty oaks and parched pastures. A comet tail of dust arose behind him as he tried to outrace the storm to his house. Before long, he heard from behind a stampede of raindrops that belied the dry road and brighter skies in front of him. He kept his eyes mainly on the road and his manic feet, because it seemed that by looking down he could make the bike go faster. When he finally looked up and saw a white shard in the distance that was the winter house, he felt reassured. Suddenly, a gust of wind from behind caught him like an unfurled sail and thrust him forward just as the heavens opened up. Instantly, he was drenched and nearly blinded by the fury of huge drops that pelted his face. The wind seemed to come from all directions now and slowed him to what felt like a standstill. Only when he fixed on a bleary tree beside the road could he tell that the frantic work of his wiry legs was doing anything at all. At times the front tire would disappear with a sudden jar in one of the rut-filled puddles, which until then, he eluded as if he were on a slalom course. He looked down most of the time to avoid the stinging drops, but when once he did dare to look up, he saw her. She was standing at the head of the cinder drive, impassive and soaked, covered by a lightweight shawl. She did not have an umbrella or a cup of hot chocolate or a shielding poncho. She wore neither concern nor anger. She was there like a harbor in a storm, like God had put her there. Only his mother would stand planted in the rain, planted, because she knew her presence alone, the mere sight of her, would be for her child, like a glimpse of the promised land.

Some years earlier, on a mild summer day, Symie did briefly disappear from his mother's sight, an episode too early on to make an impression on Symie, but distressing enough to make a lasting one on Hinda. During the summer months, the Rezniks collected rainwater from the roof of the winter house. The precious water washed down the asphalt roof into the gutter and down a drainpipe into a collecting barrel. Softer than well water, the girls used it for bathing and Hinda used it to wash the finer clothes. If some was left over, she used it to wash the overalls because it required less soap. On this summer day, the barrel collected more than rainwater. Before Symie could swim, his curiosity caused an incident that taught him respect for even small bodies of water.

While Hinda was preoccupied with the passage of a group of gypsies in front of the farm, she lost track of her youngest son. As she watched the families pass by loaded with young children, she lost herself in the thought that Symie would probably be her last child. Though he was nowhere in sight, she did not worry because the farm was a safe place to run free for a young boy. She allowed herself the brief luxury of meandering through the pivotal events in her past and of looking into some of the inevitabilities of the future.

Finally, as the gypsy families were almost out of earshot, her thoughts returned to the present and to Symie. He still was nowhere in sight. Alarmed now, she enlisted the rest of the family to look for him. It was Meyer who found him floating face down and limp in the rain barrel. During the moment that Symie sputtered back to life, his mother envisioned herself watching the world through a black veil followed by a lifetime of daily grieving. But Symie had generations to spend.

Besides a bike, Symie always wanted a dog, a useful creature, almost a necessity around a farm. Fate conspired to give him the only survivor of a litter of pups born on the deck of a fishing boat just as the day's catch was hauled in. The tug belonged to Al Baglow and Dave Fischer; the bitch who bore the pups belonged to the latter. Confounded about what to do with the fresh litter of puppies born right on the deck of the boat in the middle of a flurry of work, he sought the advice of his commercial partner. "Better throw them in the lake before the mother gets a good look and sniff of them." It seemed like the only thing to do, not needing any more dogs and not knowing anyone who wanted one. They were gathering up the pups still slimed with blood, and preparing to dispose of them when Al Baglow remembered a small boy who had been asking for a puppy. They chose the biggest one, and gave it back to its mother to nurse. When it was six weeks old, they gave it to Symie who named the dog Rover.

Rover grew up as black as snow is white. He was friendly by day, watchful by night, and an outdoor dog all the time. He lived in a barrel turned on its side with an entrance on the south side and a bed of six to eight inches of straw.   Most nights, he slept peacefully, because he was not easily or regularly alarmed. One early spring night, he barked incessantly for hours. He seemed deaf to all entreaties to quell his bark. Joe and Meyer went to all the windows of the winter house, but the only thing amiss seemed to be the bark of the dog. Though winter still had a grip on spring, Abe and Joe, armed with kerosene lamps, decided to have a look outdoors. They did not see Rover nor did they see anything suspicious around the house or barn. They decided to return to the warmth of their beds and complete their investigation in the morning.

They found Rover the next morning in his barrel unharmed and worn out from the previous night's baffling behavior. They followed Rover's tracks in the spring mud to a plow that was parked west of the barn. The plow had made the journey with the Rezniks from Dekalb, Illinois. In all its experiences, however, it never had a visitor like the one it had the night before. Heading southwest from the plow was a set of tracks different than the canine tracks that approached it from the east from Rover's barrel. They were larger, further apart, and indented deeper into the mud, the broad curved foot of a man-beast with five toes but with claws distinctly digging in ahead of the toes. The footprints belonged to a black bear. They were known in those parts but sightings were rare since lumbering dismantled their neighborhoods. The neighbors figured that a bear fresh out of hibernation must have wandered along the Black River until it picked up Abe's scent in the southwest corner of his land where he had worked earlier that day, and followed the scent back to its source, the plow west of the barn. Finding only a loud barking dog for its trouble, it found its way back to its winter lair hoping it had all been a bad dream. In any case, Rover was pardoned for his behavior the night before because he was greeting an unusual guest in the only language he knew.

If discovering bear tracks was an unusual encounter with the neighboring wild life, taking advantage of the bounty of fish in the area was a common one. During the unfrozen seasons of the year, Fanny and Symie, the most avid fishermen in the Reznik family, took off for the bayous, the seemingly bottomless pools of water on the locked-in forty-acre parcel of land south of the Reznik farm. The deep pools, surrounded by dense virgin forest of maples, birch, beech, oak, elm, and hickory, were the dead ends of mini branches off the south branch of the Black River. From the still watery blackness, they heard faintly the whirring and screeching of Ira Newton's milling saw filtering through the dense forest. Most days, they left after the day's chores were completed, but at times, they snuck off when their father left the farm on an extended errand. Most days, they would return with an ample haul, enough for one meal and a couple for the icehouse. But one day, supplied with only three worms, the fish seemed to jump on the poles, and they returned with close to one hundred fish. Unprepared for such good fortune, they had no twine to run the fish. So Symie fashioned a strong thin twig to serve the purpose, running it through the gills of the staggering batch of blue gills and bass. In the solitude and stillness of these pools, hidden from the unending responsibilities of farm-resort life by venerable trunks and summer foliage, Fanny and Symie fortified their sibling bonds. At ease together, they basked in the whims of the fishing gods above.

In most ways, Symie lead a conventional life for a rural boy growing up during the second and third decades of the twentieth century. After the age of six, compulsory schooling occupied most days of three seasons of the year. Symie's first eight years of education took place at Chamber School, a one-room country schoolhouse. One teacher taught students of all ages. Early on, his sisters transported him in a sled, Fanny pulling in front and Kate pushing from the rear. On the coldest days, the younger children sat in the squattest desks placed closest to the old potbelly stove that required frequent meals of wood or coal.

During his grade-school years, Symie could count on two friends, Ben Newton and Kenneth Dawe. Ben lived in a crowded one-room cabin in a locked-in wooded forty acre parcel of land contiguous with the southern boundary of the Reznik farm. He lived in a cabin with a hard-packed dirt floor that was swept daily by a bristle broom made of bound willow branches. The Newtons had talent in their hands – the mister lit up the cabin's four walls with a country fiddle and one of the sons was a master gear maker, using a small stove the size of an indoor wood bin to melt his metal before pouring it into its mold.

Kenneth Dawe, the son of the local postmaster, lived just east of the Reznik farm in Kibby. He was a "perfect friend" to Symie. At the most fundamental level, he treated Symie with respect, innocent of any Jew-Gentile gulf. Together, they enacted the fantasies of the day, taking turns playing the poor man's hero, Robin Hood. Then they played the ageless game of hide-and-seek. When Symie invited Kenny for sleepovers, the Rezniks shared their scant supply of meat with him. For one night, he was part of the family.

When Symie started high school, Kenny remained his friend, sitting beside him in many of his classes. By that time, Symie had long before lost the company of his older sisters and made a solitary daily trek to school. Before the sun rose each morning, while the farming adults of the world were plodding through their morning chores, he set out on foot for the high school located on Broad Street in South Haven exactly 3.1 miles away. Most mornings, Kenny caught a ride to town with his father who had to get an early start on his mail delivery. On many frigid or snowy mornings in

January, Symie broke a new trail along the side of an impassable Phoenix road that led to town. If a fresh blanket of snow had not snuffed the previous day's track, then a howling wind out of the northwest did. Some days Symie arrived at school a veritable snowman when the wall of a Lake Michigan squall plastered his fur cap and mummy-wrapped face with snow. No matter when he arrived, the warmth of his abiding friend awaited him.

During his latter years of high school, an incident occurred when Symie's brother, Meyer, literally went the extra mile for him. It was prom time and Symie had a date. Symie thought that he would make a favorable impression if he took his date to the dance in grand style in the family's Studebaker. Decisions about using the car were strictly in the domain of his oldest brother, Joe. Symie, recognizing he could use all the help he could get, attempted to draw his brother, Meyer, into his efforts. He asked, "Do you think he'll let me take the car?" In time, he successfully allied himself with Meyer to work on Joe. However, whenever Symie put the question directly to his brother, Joe, he was rebuked with the usual answer, "You little shit, go away and don't bother me." Symie, still hoping to sway the final verdict, worked like a demon around the farm, doing his chores and his brothers' chores, generally doing a first-class job of brown-nosing. In the end, his extra effort worked its magic. The Studebaker, spacious enough for a family of eight, curtain-free and open, only a windshield to preserve an afternoon spent on a hairdo, was his for the evening.

Symie set out in his best trousers, a jacket, and a dark silk tie. The evening was dry and mild, spring comfortable, as the Studebaker bounded along the rutted roads with its cargo, two teenagers with dry throats. The family vehicle arrived without incident at the high school. There, Symie was satisfied to be part of the spring tradition, without gracing the dance floor. The evening wore on to the next event, the traditional drive to Deerlick, a dead-end street that ran up to the mouth of Deerlick Creek where it emptied into Lake Michigan. An unpaved trail surrounded by virgin forest that hid the home of a bygone lumber baron, it was secluded and romantic; it earned its designation as the local Lover's Lane. Symie drove his date there and endured the required awkward, speechless period before relaxing with a few sighing breaths. While the laughter and shouts of his classmates rang in the darkness broken by the beams of couples coming and going, he left for home. He had done his duty as a teenage boy coming of age under the scrutiny of his classmates. He left in time to make his curfew at 11:30. When he arrived home, expecting to find everyone actively involved in their daily regimen of sleep, his brother Meyer was no where to be found.

Earlier in the evening, after Symie had taken off with the Studebaker, Meyer had wandered into the bedroom where he found Joe reading. "Hey," Joe said, "Did you tell the Little Shit about the soft tire?" Joe's job had been only to give his permission; it was Meyer's responsibility to make sure the car was road-worthy. A soft tire on the late spring rutted roads could easily have negated the favorable impression Symie wished to make by driving the Studebaker. Meyer hustled double-time the three miles into town and changed the tire while his younger brother was trying to decide whether dancing was a frivolous activity. Changing a tire in 1925 required the same tenacity and energy as pulling a stubborn jackass a couple of times around the block. By the time he finished the job, Meyer was grimy and tired and did not relish the idea of walking the three miles back to the farm on a moonless night. Instead, he climbed into the back of the Studebaker and assumed a recumbent pose on the floor and slept through the rest of the prom, the short journey to Lover's Lane, and the ride home. When Si arrived home, his brother was not far behind. He was in the back seat sound asleep from the exertion of his good deed and the late hour. To this day, the story stands that Meyer did not feel a bump in the road or hear a single awkward word. God delivered a sound sleep to the young man at the right time or endowed him with an extremely tight set of lips.

An Intended Day of Rest

When Abe swept his intended into his arms and carried her across the mud puddle, he initiated a sequence of events that took Hinda, the marginally spoiled only girl in a family of eleven, to a new land where privilege was too often elusive bait. As it was in the lives of many Jews who lived by old-world-ways in the new world, the Sabbath still held its holy place. Preparation for it was synonymous with cooking and cleaning, especially the latter. On her hands and knees, each Friday, with a scrub brush and Fels Naphtha Soap, she meticulously scrubbed from the raised grain of the pine floors a week's worth of accumulated farm life, mostly sandy soil scented lightly with manure. When she completed the task to her satisfaction, she covered the floor with newspaper to preserve its state of cleanliness for the evening. Finishing at the doorway of the kitchen, she stood up and, to no one in particular, said, "Ahbeegezunt", meaning "as long as we are well."

By noon, a loaf of white challah was cooling on the rack. This white egg bread was special because it was made entirely from finely ground flour instead of the more plentiful coarse cracked wheat flour. The Sabbath brought a departure from the usual dairy milchig menu. During the winter, Hinda either used a crowbar to pry free a quarter of the frozen geese parts from the half-buried barrels or she prepared a portion of the butchered calf or cow. The men and boys returned from their chores earlier in the day on Fridays to prepare for the Sabbath. Discordant aromas of cinnamon rolls and homemade gefilte fish greeted them. They washed the day's toil from their hands and faces, dressed in their finest trousers reserved for the

Sabbath, and covered their heads with skullcaps. All the Rezniks then assembled in the kitchen where the candles were lit and the blessings made. Abe stood up with his hands stretched toward the heavens in front of him and motioned, as adherent Jews had for thousands of years, for God to enter his house.

Abe observed the Sabbath by departing from the usual time that he assembled his fifty cigarettes, doing it before supper rather than after when the sun had set. At first, he also refrained from smoking as well, believing that smoking tainted the holiness of the day. But smoking had become a chain for him that could not tolerate a missed link. At first, he was secretive about it, smoking with the fruits and vegetables in the cool, musty Michigan basement. As the entry was not airtight, Hinda soon identified the unusual odor emanating from the depths of the house. For some months, she said nothing, not wishing to amplify his shame. But she finally concluded that it was a greater shame for him to act like a second-class citizen in the home he had worked so hard to maintain; she broke her silence, "Tzalie, you don't have to smoke in the basement anymore."

Abe had to make a more profound compromise with his conscience than the one his smoking required of him. That he could not fulfill the commandment of rest on the Sabbath created a permanent pang of guilt. As his farm did not rest on the Sabbath and he could not afford to pay a gentile to do the daily chores, he was forced to supervise an abbreviated day of work to appease his livestock and poultry. For this transgression, his conscience never let him stand on the pulpit during services. He hoped his merciful God would accept this act as an ongoing sign of his repentance.

The Realization of a Dream

When the Reznik family settled in their corner of southwestern Michigan, the nearest concentration of Jews significant enough to attract the regular spiritual guidance of a Rabbi was twenty-eight miles to the south in Benton Harbor. There, the Jews could congregate on High Holidays and the Sabbath and to celebrate Jewish life's milestones such as Bar Mitzvahs, weddings, and the commemoration of loved ones. They could assemble under the roof of a structure devoted to the worship of God – a synagogue. Within its walls, the pious could come and dovan, or pray, and feel a closeness to God and to their history reminiscent of their homeland. The journey by horse-drawn wagon from the Reznik farm to this synagogue over eroded roads, a round trip of fifty-six miles, would consume eight to ten hours in preparation and travel time. It was a luxury that a struggling small-farm family could not afford. But as the Jewish Agricultural Society helped other Jews move to the South Haven vicinity, another possibility emerged for a Jew who lost the anchor of his synagogue to gain a measure of social and religious freedom. The dream of a synagogue built among the Jews in rural South Haven ignited in Abe.

The fulfillment of his dream would take time and money, time clocked in years because the money required was measured in units of thousands of dollars, an extraordinary sum for Jews who could barely make ends meet on their small farms. While this dream was incubating, the gravity of the High Holidays annually exerted their ancient hold on the Jewish families around South Haven. Their observance, before the construction of the synagogue, took place at the Reznik homestead during the warmer days of early autumn. After the racehorse barn had been converted to a large dining room with a second floor of guest rooms, the Reznik farm became an ideal location for Jews to congregate for dining on kosher meals, for worshipping, and for fasting. As many as fifteen families, from a twenty-mile radius, congregated at the Rezniks. All of them were observant orthodox Jews, forbidden to travel after sunset on the first day of the holiday until after the sunset two days later. For Rosh Hashannah and Yom Kippur, the families, therefore, remained guests of the Rezniks for three nights and two days. For the duration of the holiday, each family left their wagon in the barnyard and put their horses out to pasture in the fallow Reznik fields. Each family was afforded one of the resort rooms, but was responsible for bringing their own meals. The food, prepared prior to the holiday, fish, potatoes, dairy products and sparse amounts of kosher beef, was stored in the icehouse.

The friendly service available during the resort season, naturally, was not available during the period of the High Holidays. Each family took care of itself. One older couple, the Gassins, occupied a pair of seats closest to the exit of the dining room to minimize the trek necessary to retrieve food from the icehouse. Mrs. Gassin, crippled by years of honest labor, hunched over her meals like a sapling under the weight of a heavy snow, immobile except for the awkward, tremulous movement of her hand from the plate to her mouth. Her doddering husband who moved like the second hand of a clock, made his way back and forth from the icehouse, and served his wife before arranging his own meal. His devotion suggested a flame that once burned very bright for his wife and for his God, but had subsided to a warm gray bed of coals.

The services were a solemn all-day affair with no inclination to abridge any service. Meals, sleep, or rest divided the day into its different services. They were conducted most often by one of the Rabbis that lived in the area or by a pious community member familiar with the service from the years of repetition, but the Rabbi always gave the sermon. The restless children, on whom the droning and dovening and the solemnity of the day were lost, periodically were freed to play outside long enough to consume their fidgeting energy. By the conclusion of the service, the two-day retreat at the Reznik farm satisfied the hollow place left by the absence of a synagogue and tradition from the old country. The Rezniks, who opened the gates of their homestead, were no less the beneficiary of their mitzvah than any other family. Under the circumstances, no money ever exchanged hands.

Raising money for the shul began early on in the formation of the Jewish farm community. Abe, who had been among the first of the Jewish farmers and who had in small ways aided the migration of other Jewish families to the area, took a leadership role. He helped raise money for the shul during a time when a spare dime was as scarce as a 60-degree day in the dead of winter. The pancake dinner became the principle means of raising money. During the winter, when the resorts were a warm memory, each month a Jewish household prepared a pancake dinner. The other Jewish families, committed to the idea of a rural synagogue, congregated at this house and filled themselves with pancakes. For each pancake eaten, a one-cent contribution was made toward the construction of a synagogue.

The sentiment of this endeavor contrasted with the hardships of achieving it. At the time these dinners occurred, the Jewish community was spread out over a ten to twenty-mile radius. During the dead of winter, when the sun was already low in the sky by the time lunch was digested, each family contemplated a one or two-hour lantern-lit trip by sleigh through frigid night air whipped by strong westerly winds. If they were lucky, the snow developed a hard crust; but if it was deep and fresh, the going was more cumbersome yet. These frigid nights would have been intolerable if it were not for a fire keeping them warm – the commitment to a synagogue, the anchor of the shtetl, deep-seated and fundamental.

The Rezniks had to prepare for a trip like this. The horses, the biologic engines for the affair were protected from the slicing, frigid winds by a couple layers of horse blankets belted beneath their harness. For themselves, they first filled the sleigh with a bed of fresh dry straw. For more warmth, shortly before leaving, Hinda heated large stones on the stove, wrapped them in newspaper, and buried them in the bed of straw. Then, like fingers in a mitten, the children and their mother squeezed into the back of the sleigh. They covered themselves with horse blankets from the soles of their feet to the knolls of their chins. Symie, the little one, usually submerged himself completely in the mass of warm human bodies beneath the blankets, coming up for an occasional snort of the icy night and a glimpse of the emerging myriad of winking stars.

Over a period of more than six winters, the Reznik and like-minded Jewish families endured winter's vicissitudes for a meal of pancakes. The endeavor symbolized both the spirit and the hardship of building a synagogue in a rural community still remote from some 20th century amenities such as paved roads, electricity, and indoor plumbing, but a community instilled with a strong work ethic and a yearning to rebuild the center of its old universe.

The community generated other sources of money to fund the hope of a synagogue. At times, summer resorters became aware of this grass-roots effort and generously contributed five or ten dollars. A more important source of revenue, however, came from the schecting work, the sacred slaughtering, done by Rabbi Schwartz, a beardless non-orthodox orthodox Rabbi, who in spite of his profound doubts about the existence of God, assiduously followed all the kosher laws. In spite of his spiritual contradictions, Rabbi Schwartz, who had enough strength to schect a cow as well as a chicken, fulfilled the need for a schochet in the community. If he schected a cow, for his trouble, he received some of its beef, but if he butchered a chicken or a goose, on account of the chasm which separated him from his God (though not from his history and brethren), he would not accept payment for his work. Instead, the Rabbi kept a detailed calendar recording all the poultry he butchered and for whom he did it. At the time of service, no money was collected, but at the end of the season, Abe used the information on the calendar to collect a nickel for each chicken schected and a quarter for each goose schected as a contribution to the synagogue fund. During that week, Abe's digestive system went through hell as he tried to collect these hard-earned buffaloes and standing liberties from families who, during certain times of the year, could not spare a Lincoln copper.

Abe accepted the responsibility for safe keeping all the money collected for the shul. He collected what seemed a miraculous amount, but without one more substantial, generous contribution, the synagogue, in all likelihood, would have remained a dream. In the end,

God intervened in the form of a five thousand dollar donation from the CEO of Sears and Roebuck, Julius Rosenthal. Five thousand dollars was an extraordinary amount of money in the days after the Great War, especially to farmers who were strangers to luxury. The Reznik farm-resort which prospered as much as any of the neighboring Jewish farms generated only five hundred dollars during a productive season, most of which was consumed in providing for the necessities of daily living during the winter and start-up in the spring. If ten or twenty dollars remained after covering all the necessities, and twenty Jewish farms were this fortunate over a succession of years, it would have taken between ten to fifteen years to generate five thousand dollars. To generate this sum, Rabbi Schwartz would have had to butcher twenty thousand geese or one hundred thousand chickens. In 1920, five thousand dollars would have purchased two brand new Cadillacs, or thirty-three Ford Model T's, or forty-five phonographs, or twelve thousand pounds of chicken. For Abe and other yearning Jews of his community, the donation was a God-send, manna from heaven, a mitzvah gadolah, a dream-maker.

What began as an indistinct possibility had become a palpable probability. Though everyone shared the dream of praying in a local synagogue, not everyone agreed on where it should be built. Most of the Jewish families of deep faith but shallow means lived on farms scattered around the countryside well east of the town of South Haven. These families were exposed to the modified breezes of the lake but not to the view of its waves and ripples. These families, in the majority, favored a shul in the country for people who lived in the country. A vocal minority, however, who lived in South Haven and ran prosperous resorts that benefited from electricity and indoor plumbing, favored the construction of the shul in South Haven proper. Two camps formed, one "have" and one "have less", one rural and one town.

The matter was not settled without acrimony. During one of the heated debates in which men in skullcaps hurled vehement Yiddish phrases at each other, blows were exchanged, Jew against Jew. The worst of these blows landed around the right eye of the shoemaker, Sanderovich, apparently knocking the eye clean from its socket; those who could stomach it thought they saw it dangling there. For weeks, Sanderovich made the weekly trip on foot to town so that Old Doc Penoyer could have a look. If Symie saw him trudging by and nothing was pressing on the farm, he would give him a ride into town. In the end, the matter was settled, in large part, because the money donated by Julius Rosenwald was earmarked for a rural synagogue for farmers. With his impaired depth perception, Sanderovich watched the construction of the synagogue three miles east of South Haven at the intersection of two unpaved country roads.

The synagogue was built on a donated plot of land a quarter of a mile east of the Reznik farm, adjacent to the Fidelman farm. Mr. Janosky, a local gentile builder known for his honest dealing with the Jewish community and for his craftsmanship, was commissioned to design and build the synagogue. He erected a simple structure with a gray stucco exterior and clean white interior walls. A meticulously constructed arched ceiling swept upwards from twelve-foot walls. Ample windows flooded the chapel with light from a rising or setting sun, casting golden and rose hues on the opposite white walls. Unvarnished oak planks, abused by the gathering of pious and joyous feet, would require frequent bouts of sanding to remove the stains of future spills. An unfinished full basement with a cement floor remained vacant with the exception of a stove, a kitchen and a toilet. The completion of the basement awaited more opulent days and an expanding membership. In all, the structure was spacious for twenty Jewish families and its' simplicity was harmonious with the fiber of the people whose dreams it fulfilled.

The ark, built by a Jewish artisan living in Paw Paw, was erected on the eastern side of the shul. The sacred receptacle of the holy scrolls spanned floor to ceiling. It was a ruby set on a simple band, fit for a synagogue supported by a base of fervor rather than wealth. With the ark to its rear, the bimah stood in the center of the shul overlooking neatly arranged rows of oak folding chairs. It was a platform of solid oak construction rising out of the sea of oak below, large enough to handle any occasion-- the solitary sermonizer of the High Holidays, the chupa or wedding canopy of a wedding, or the proud family of a Bar Mitzvah, a Jewish boy who became an adult member of his community. The bimah and the ark embodied the expectations for the shul's future.

The shul had a roof, an ark, a bimah, a small congregation, but did it have a Rabbi? The answer to that question is an unequivocal, "Well, yes and no." The country around South Haven became the haven for the aforementioned brilliant, but non-traditional Rabbi Schwartz, a flawed jewel cut among the scholars in Europe. Like many Jewish settlers in the Kibbie area, the Rabbi initially experienced the southwest corner of Michigan as a guest in the Reznik home through its connection with the Jewish Agricultural Society. He and

Abe formed strong bonds early on that lasted until the latter's death.

Unlike his fellow orthodox Rabbis, Rabbi Schwartz was clean-shaven. Sometime during his education, he had become an atheist, but not before he adopted many of the traditions of Orthodox Judaism. From the start, he was a natural contrarian, particularly taking the opposite point of view in any philosophical discussion. Before immigrating to the United States, he developed a reputation as a skilled debater among the rabbinical councils in Europe. He made the rounds in Europe, skillfully arguing any point of Talmud if it did not cause him to strike down the central supporting beam in his own soul, that there could be good Jewish people, laws, and traditions, without a God. He would eventually become a thorn in the side of his bearded brethren, and he was forced to look across the Atlantic. During the time he passed from a reticent atheist to a more outspoken one, he did pass the necessary exams to become Ruff, a Rabbi capable of leading his own synagogue. Though he rarely spoke of it, before he departed Europe, he had led a sizeable congregation and had ascended the rabbinical hierarchy to a level comparable to a cardinal in the Catholic Church.

When Symie knew him around the time the synagogue was built, he was in his early fifties, just beginning to gray and a few inches taller than his father who stood around five feet seven. A very lean man, he was regularly wracked by the cough of a chronic lung condition. Not only did Abe share companionship with the Rabbi and consider him his bosom friend, but also he shared this propensity to cough. When the Rabbi conducted a service, he regularly required relief after an hour, too winded to continue when his natural animation took its toll on his chronic lung condition. He would therefore save his strength only for the High Holiday services of Yom Kippur and Rosh Hashanah, choosing topics for his sermons that were both non-traditional and at the same time pertinent to the lives of his hard-working farming congregation. In his speeches and sermons, he was able to penetrate a world of ideas that this community might otherwise have never known. His only rival in brilliance and leadership and skills as an orator was the young Rabbi Cohen who was precocious and wise beyond his years, ordained as a Rabbi before starting his third decade of life. When these two mental titans matched wits, the veteran atheist and the youthful wizard, onlookers like the young Symie were both awed and entertained.

Children took to Rabbi Schwartz like he was a rabbinical pied piper. He entranced them with stories and intricacies from the Talmud, filling these with suspense and elucidating the issues relevant to their lives. He seemed a river of knowledge, easily and quickly absorbing any stream or tributary of new facts or ideas. He was fluent in English as well as many other languages from the motherland. Yet, as beloved as he was, as generous as he was, as knowledgeable as he was, for certain things, the parents of these children did not trust the Rabbi. After all, he was a professed atheist. Could they entrust their children to a Rabbi who did not believe he was a man of God even if his deeds were evidence of His existence? Every Jewish boy raised in an orthodox family considers conducting part of a Sabbath service when he becomes a Bar Mitzvah at the age of thirteen. To chant a portion of the Torah, he requires hours of instruction from a teacher like the Rabbi. To turn to a generous man like Rabbi Schwartz for instruction should have been natural; the next closest suitable person, depending on the mode of travel and the weather conditions, was several hours away. But at a fundamental level, the community perceived his atheism as an infectious disease. It could be transmitted to a child if he were in close contact long enough without a parent to sanitize the insidious germ.

If the circumstances of the Rabbi's life somehow caused him to gravitate toward a belief in the existence of God, a single event negated any movement forward. If somehow his central role in a rural Jewish community allowed him to feel whole again, having a hand in the death of his son left him a broken man, cracked inside like the outside of a raku figurine. His sixteen-year-old son fell ill with a pain in his abdomen. Inexplicably, the Rabbi ignored his son's complaints. Not a pious man, he certainly did not put his son's fate in the hands of God, even if it really was. He also ignored the exhortations of his neighbors to seek medical attention for his son. By the time he made his move, his son was fatally ill with a ruptured appendix. It was a circumstance from which no conscientious man could be expected to recover. He had no God from Whom he could beseech forgiveness, and he could not find his way to bestow it upon himself. The event extinguished his enthusiasm to argue the other side of any issue. How could he care about words and ideas when he had lost the dearest tangible thing in life? The shame and guilt consumed his brilliant mind and caused him to recede from the other lives in his community.

The last event in which Si recalls Rabbi Schwartz's participation was his bosom friend's funeral. His active role in Abe Reznik's funeral was a final testament to the close ties the two coughing men had developed during the formative years of the rural Jewish community. Only the need to offer comforting words and a deserving eulogy could have mobilized the

Rabbi, who made his living as an optician, from the paralysis of his failure. With the death of the Reznik patriarch, however, went the final reason he had to visit the Reznik household. The butchering of aged cattle had long since ceased with the establishment of a kosher butcher in town in the early twenties. There was little call for his precision in schecting poultry. His last days flickered and faded under a pall of contradictions, seldom called upon to shape a set of lenses, searching for the pair he lost long ago through which he once saw meaning and purpose in life.

In too many ways, the history of the First Hebrew Congregation, the rural South Haven synagogue, paralleled the life of one of its first contributing Rabbis, Rabbi Schwartz. It began as the culmination of the religious aspirations of a scattered group of poor Jewish farmers, a symbol of what other distant dreams they might realize in their lives. But the beauty of its beginning turned out to be the sunset, not a sunrise.

There was an auspicious beginning though. Before the carpenters withdrew their tools and swept up their wood scraps, the shul precociously witnessed its first wedding. In 1922, Anna Reznik offered her vows to her groom. From a twenty-mile radius, Jews congregated to be part of this inaugurating event. In preparation, the congregants started in the center of the hall and pushed the sawdust, wood scraps, wood chips, levels, hammer, saw horses and nails over the freshly installed oak floors to the edges of the room. Then they set up oak chairs in view of the central chupa that was adorned with white silk. Rabbi Pearlman, an aging, frail, traditionally bearded Rabbi, the counterbalance to a room permeated by the newness of fresh-cut oak, presided at the ceremony. They celebrated that night with the sanguine belief that this ceremony would be the first among many as life would repeatedly cycle to this celebratory moment. But it was the first and the last wedding in the Holy Hebrew Congregation.

While the construction of the synagogue was an implausible feat akin to making a small amount of oil last a few days longer than expected, the seeds of its demise were sown at its inception. One might understand the short existence of the synagogue by comparing its chances of success with those of a corpulent person vowing to achieve slimness the rest of his life. A middle-aged father with a protuberant belly hanging over his belt might decide he is going to drop sixty pounds for his daughter's wedding so that he can fit into the tuxedo in which he was married. He finds inspiration in his love for his daughter and a sincere desire to provide a symbolic continuity with his own wedding ceremony. He adopts a Spartan diet with a third of the calories of his usual affluent diet. Furthermore, he burns another five hundred calories every day jogging in all conditions except tornado weather. With his new adopted habits, he experiences an infusion of energy and he is prolific in all the spheres of his life. In spite of his super human effort, the wedding is two weeks away, and he still has ten pounds to lose at a time when the rate of weight loss is leveling off. When, in a state of discouragement, he is about to reach for a piece of cake and skip a run, he awakens the next morning achy, chilled and nauseated. For three straight days, he cannot keep anything down and for the next week, everything he eats seems to run right through him. When everyone around him is despairing that he won't be able to stand at his daughter's side, he makes a miraculous recovery and gives his daughter away, wearing comfortably the tuxedo in which he was married. As his daughter gives her vows, he makes one to himself and to his future grandchildren to keep the weight off. But he has expended his allotted Spartan discipline for that decade. He slowly slips back into the habits that fit like a worn shoe. A year later, he wears his belt buckled one hole beyond the one that gaped with wear at the time he started his diet.

In the same way that the motivated loving father set about losing weight, the inspired members of the community set about raising money for the synagogue. Their efforts were achieving a crawling success; but, just as it is uncertain whether our corpulent father would have succeeded without his fortuitous illness, it is unclear whether they would have succeeded in their efforts without the final infusion of funds from Julius Rosenwald. Once the synagogue was built, using and maintaining the synagogue was no less an ongoing job than raising the money in the first place, an endeavor reminiscent of the father's attempt to keep his weight off. But no degree of dedication could make up for the fact that the Jewish rural farm families were still few in number and not well off with little free time on their hands.

The synagogue was still ten miles away from some of its supporters who, under the best of circumstances, had to spend half the day getting there and back. Under the worst of circumstances, when a biting wind sculpted the unplowed, unpaved roads with waist-high drifts, the synagogue was inaccessible for days at a time. After days or weeks of vacancy, the heat from the previous service had long since escaped; and it became necessary for Abe to get an early start on making the sanctuary habitable. If a long service was anticipated, he fired up the pot-bellied stove with coal; for a shorter service of three to four hours, he used wood. But how long could a building built on lofty aspirations survive on the backs of a few?

The small congregation used the synagogue irregularly with the exception of during the Jewish holidays. Sabbath services occurred sporadically, mainly when there was a Yahrzeit, the anniversary of the death of a loved one. The male member of the congregation most knowledgeable in Hebrew and the Sidur, the Jewish prayer book, led the service. At times, it was still more convenient to hold a service at someone's home. Dovening in someone's home averted the need to warm the shul from frigid to tolerable for an impractically short time. Services were held more predictably in the shul on the more important Jewish holidays. The High Holiday services were universally attended and were led by a Rabbi. Other services took place on Pesach, Shavueth, Chanukah and Purim.

The High Holidays annually rejuvenated hope for the synagogue as they started a season in which it received regular use. But with the onslaught of frigid winter days and then of the daily tasks of farm-resort life, the synagogue slid into the background. Soon, in the way the dieting father's belly overlapped his beltline, the synagogue became unkempt, no longer a fitting centerpiece for the South Haven Jewish community. The voices from town that had previously been quelled by the rural missionary zeal became greater in number and more insistent, in English; and what's more, they promised gelt. Their resorts in town, close to the lake, the train, and the steamships, outfitted with indoor plumbing, sewer and electricity, attained an elegance and comfort that quickly made the farm-resorts a thing of the past. Why should they travel three miles over tire-eating roads to a synagogue that was cold and had peeling paint and stained, unvarnished floors, and oak chairs instead of pews? Why travel to a synagogue that had the feel of a farmhouse instead of an urban mansion? The resorts in town were flourishing while the farm-resorts were declining or closing.

The rural synagogue never made it out of its infancy. It was abandoned after only five years for a bricked second First Hebrew Congregation in town dedicated in 1928, with oak pews, plum carpet, and a balcony for women and children. The nourishing spirit of the country synagogue withered away and was supplanted by a more sophisticated Chicagoan one that was one more generation removed from the shtetl. Like the loving father who once stood proudly by his daughter in his black tuxedo but succumbed to a fatal fatty clot in his heart before he saw his grandchildren reach kindergarten, the synagogue, without its inspiration born of memories of the shul in the old country, succumbed to a fatal spark and burned to the ground.

The End of an Era:
The Reznik Farm-Resort Closes

For Symie Reznik, the farm-resort three miles east of Lake Michigan was the only world he knew intimately. The rest of the world was out there, but it did not beckon him. Nor did it call out to his mother and father who had chosen this plot of land to be their promised land. For Symie's brothers and sisters, however, there was a world beyond the forty acres where they had spent the most recent segment of their lives. They aspired not to the world of their past, the subsistence days on the farm in Sandwich, Illinois, but to the world imported by the cosmopolitan resorters from Chicago. The girls interacted with it regularly when they brought resorters homemade gefilte fish, boiled potatoes and onions, a slice of raspberry pie or whatever was in season, and a glass of ice water to wash it down. They worked hard without flirting, but they listened and observed; somewhere, among the guests, lurked the man with whom they were going to spend the rest of their lives.

The Reznik girls were bound to a farm in a community of a few dozen Jewish families scattered over a twenty-mile radius, without the services of a matchmaker. Their prospect for finding the man of their dreams seemed as unlikely as finding a robin in the dead of a harsh winter. Were it not for the resort that attracted Model T's and occasional Chevrolet 490's loaded with eligible young Jewish bachelors, the Reznik girls might have faced a dismal outlook. With this in mind, the Reznik girls, when they reached a lull in their daily chores, watched from the dining room window as the dusty automobiles arrived. They awaited the first glimpse of a patent leather shoe or a plastered sheen of dark hair above deep-set brown eyes, anticipating the disappointment of a lady in tow. The vehicles loaded exclusively with boys generated a half-playful volleying for the best view, and on one occasion, a resolute claim and prediction. Fanny, shoving between her sisters while her gaze moved between a pair of brown patent leather shoes and the promising face that complemented them, had a profound premonition. She turned to each of her sisters and said, "Lay off of him. I'm going to marry that man for his brown patent leather shoes." Charlie Pierce Pivovitz, a young Jewish man who had just driven to South Haven for a little R and R with a couple of his companions, had no idea that had he worn his wingtips, he might have returned to Chicago rejuvenated but unclaimed. As it was, he had been selected; and unbeknownst to him, the course of his life now had road signs with a destination.

Meeting Charlie took place naturally for Fanny in the course of waiting tables. Fanny, the youngest of the three Reznik sisters, was the least fettered by restrictive rules of conduct when it came to socializing with the patrons. Work and fun were not mutually exclusive enterprises. After making the connection with lively amiability, avoiding suffocating extra attention, the road signs came into focus for Charlie; and he began to take regular vacations at the Reznik resort.

Two years separated the proposal of marriage and the ceremony, for Fanny, an interminable amount of time. Among the problems with Charlie was his name. He had the misfortune of sharing the same name as Fanny's father, Abraham "Charlie". According to orthodox belief, introducing another "Charlie" into the family might bring misfortune to the original bearer of the name. If that were the only problem, perhaps Abe might have overlooked what's in a name; but Charlie, at least when Fanny got to know him, managed a pool hall in Chicago, which while lucrative enough, suggested shady characters and transactions. While these unyielding obstacles stood between courtship and marriage, Fanny was grief-stricken.

Fanny's normal cheer vanished as she mechanically completed her chores. Instead of going fishing, she spent most of her free time in bed sobbing or moping in view of her siblings and mother. One time, before events took a more positive turn, Charlie sent a five-pound box of chocolates to Fanny to assuage her grief. Under normal circumstances, since candy around the Reznik house was scarce unless it was Chanukah when each of the Reznik children received a peppermint patty on each of the eight nights, the box would have been worthy of a minor celebration or declaration of love. Fanny was not tantalized by the possibilities of this box, but her younger brothers were. While she lay up in her room in bed temporarily dry-eyed and comforted by the unopened sweet-scented box, her two youngest brothers, like a matched set of over-wound cuckoo - clocks, chimed every five minutes from the bottom of the steps, "Can we have a piece of chocolate candy?" In the end, their relentless requests wore her down. She appeared at the top of the steps and hurled the unopened five-pound box down the steps, hoping to do some damage. She yelled, "Have some chocolate you morons." To this day, Si remembers how the white chocolates in that box melted in his mouth.

The siege on Fanny's heart ended when the persistent appeals of her mother opened the gates to her father's heart. By that time, Charlie no longer managed the pool hall. He worked for a paper vender who collected and sorted paper into four different grades. They were married in Chicago, which eventually became Fanny's home. After her father softened and before she left home a married woman, Symie again conversed with her about his high school exploits on the way to the back forty-acre fishing hole.

The farm-resort was fruitful for the two older Reznik girls as well, producing a husband from Chicago for each of them. One of the husbands had accompanied Charlie Pivovitz when Fanny first saw him, but he did not elicit the same resolute claim. The older Reznik sisters, also, acquired a taste for life in Chicago and adopted it as their home away from home. Chicago, though it was not filled with jobs suited for a lady, did provide them with relatively high paying, steady jobs in the sweatshops at twelve dollars per week. During the summer, they rotated through the resort helping out until Abe died. They brought increasing numbers of children as they came along. They shacked in the blue-room beside the kitchen in the winter house where a sour hint of winter's cheese lingered on sultry summer days.

With the exception of Symie who was still in high school, the younger Reznik men also found partners for life on the farm-resort who moved the center of their world across the lake to Chicago. Their sophisticated wives presented a force stronger than the filial and nostalgic ones that rooted them on the farm-resort. What's more, in Chicago, two honest, trustworthy men raised on hard work could find more reliable, better paying work than farm-resort life offered. Besides a greater abundance of work opportunities, the west side of Chicago offered a greater concentration of Jews and synagogues spanning a broad range of liberal to orthodox beliefs. They had no trouble finding a Sabbath service with a congregation that approached their ancient faith in a similar way. The ingredients for a different life for all the Reznik children, except Symie, were assembling away from the kitchen of their formative years.

Abe Reznik, while his children's affinities drifted across the lake, paid attention to the west for another reason--on this horizon, the subsiding sun drew each day to its close and laid to rest another grain of sand on the small dune of his work life. Every day he arose, as reliably as crocus shoots sprout during the first warm days of spring, as certain as one breath follows another, to work. From a distant perspective, each day, each week, each year resembled another, a sequence of tasks created by the husbandry of a small dairy farm-resort, as monotonous as a hike through a sea of ripe grain. Yet each day, he arose relentlessly, inspired, oblivious to the repetitiveness of his daily toil, because he believed that his work formed roots that connected to an ancient tradition from which his progeny would blossom with new possibilities. So when Abe did not arise from bed one day in his fifty-sixth year, something was very wrong.

From his youthful days in the Russian army to the mature years of grandfather-hood, he had bathed his lungs in the acrid fumes of unfiltered Turkish tobacco. He adapted to its ravaging effect with a cleansing ritual each morning before work, summoning up and expectorating, before the rest of the household stirred, a night's worth of accumulated secretions. With each passing year, the ritual became more tiresome as the Turkish tobacco stole more and more of the air he needed to combat its damage. When at one end of the eight-foot bow saw he had to rest with greater frequency and when he retired from using the walking plow and used only the horse-drawn plow, he chalked it up to the same forces that make the once invulnerable youthful oak susceptible to the gusts of November storms. When finally at ease with the existence of leisure in his life and a worldview that permitted a little pampering, he bounced his grandchildren on his lap, a luxury he had never lavished on his own children. They were bewildered by the coarse, labored breathing that emanated from this indestructible man. The sound reminded them more of a winded horse than a human. The sucking air sound to which he was accustomed began to alarm everyone around him. But no one imagined that one day he would not get up.

When Abe did not get up for a second day, Dr. Penoyer was summoned. Yes, his lungs were very bad, but his kidneys were a problem as well. He was not making water and his tissues were filling with the undrained fluid, distending the wrinkles of his face, and creating a strange rather than a youthful look. On the third day, visitors were not permitted to stay more than moments as he went in and out of consciousness, struggling for his last gasps. For those who had not watched Abe overcome everything to arise with the sun each morning, they would have conceded that he would never rise again. For those who watched daily, they clung to the hope the next sunrise would resurrect past days. They beckoned him in Yiddish with images of the growing grandchildren, trips to Chicago, no more toiling in the paralyzing summer heat, the reunion of Sabbath meals, and spring rejuvenations. Rabbi Schwartz who knew too well the brand of the Angel of Death usually prayed quietly, almost silently for his friend. At intervals, he raised his voice to share a prayer and his premonition with a family member who refused to see what was right around the corner. On the third day of his lapse, Abe turned that corner and his struggle ended.

Hinda Reznik's family surrounded her during these darkest of days. They witnessed a heart-wrenching aftermath. For Hinda, Tzalie had seemed capable of anything, and one day he did not get up to work and three days later he was gone. From Vilna to London, from Sandwich to South Haven, they had welcomed the morning together, pursuing a dream of land ownership, the anchor of the schtetl-the shul, and a life filled with fewer hardships for their children. For their God and family, they had endured anything, and now she would have to do it without him. The wound was fresh and gaping at the funeral. Her brothers from Aurora, Illinois, who came to console her, could not restrain her. She flung herself over the coffin, her husband's visage shrouded forever behind a wall of oak, and sobbed and howled with grief. When she found words, she repeated them like a mantra to ease the pain that flooded her soul, "Tzalie, du szolst betten gut far unzeva kinda (Charlie, you should beg God to take care of our children)". There was no consolation for her grief; he who had arisen with the sun and outshined it would never rise again. She was permeated with unadulterated anguish, submerged alone in the sunless depths of a gulf of pain the size of Lake Michigan with no choice but to inhale it into her lungs where it circulated to the fingertips of her being. When her agony reached an intensity that resonated with and summoned the howls of loss from the beginning of time, filling the synagogue with countless ghosts, her brothers carried her away.

For the community at large, Abe's funeral was a noteworthy event. In the estimation of some, it marked the passing of a founder or a pioneer, and by all accounts, the passing of an honest man, Honest Abe. His was the only funeral to take place at the countryside synagogue. After embalming and before closing his coffin, he lay for a day on a palate of straw for visitation. After the funeral, a procession of buggies and wagons one mile long formed behind a black hearse drawn by two black horses carrying the first American patriarch of the Reznik family. The asundry vehicles proceeded from Lacota to Bangor where the coffin was placed on a train bound for Chicago. There, he was laid to rest in a Jewish cemetery. On his last day in western Michigan, Jews and Gentiles alike remembered him. That day was the beginning of the last chapter of the Jewish farm-resorts.

For a week after the funeral, Hinda sat Shiva in a numb trance. Her daughters, as they had done throughout their lives, attended to the visitors who brought traditional meals of rolls, bagels, and hard-boiled eggs. According to orthodox custom and consonant with her heart, she did not listen to music or attend a simcha, a party, for one year. Yet, she recovered to perform her solo role at the helm of her family and what was at one time perceived to be its legacy, the farm-resort.

When Abe died in 1926, of the Reznik children, only Symie remained in South Haven year round. He still had two more years of high school to complete. Symie along with his mother, the hired help, and the summer assistance from the other Reznik children, could no longer adequately farm the land or operate the farm-resort. Abe had been the spiritual glue of the farm and its most important manual hand; the farm would not survive the absence of both. Only so that Symie could finish high school in South Haven did Hinda remain on the farm-resort. For his sake, she stayed to witness the final decline of the first First Hebrew Congregation as the movement to build the synagogue in town gained momentum. For his sake, she stayed to witness nature's haphazard reclamation of the once neatly plowed fields. She received frequent visitors, some from Chicago and others from the community, which created snatches of the former comfort and familiarity of her home. But something fundamental was missing, not only Abe's daily departure and return with the cycle of the sun, but something almost as tangible, Abe's dream - the one that had been fulfilled when they bought the land and later built the synagogue. At the same time, the dream had been spent and had borne its fruit. For Hinda, the end of the farm-resort and the synagogue were inevitable. Her children had grown up free, with a strong work ethic, with the possibility that life would deliver a more penetrable rind than what she had encountered.

The tourists from Chicago, Detroit, St. Louis and Ohio, who once slept beneath the Reznik fruit trees while they awaited the July 4th sunrise, ignored the country resorts by the late 1920's for all the 20th century conveniences in town. If it were not for the proliferation and mobility of automobiles, the quiet countryside would have been even quieter, almost vacant in places. By 1928, the Reznik resort, already short of help, was starved for income. Hinda could no longer afford the payments on the farm; the farm passed through the hands of the bank back into those of its original owner, Bill Trowbridge. Hinda and Symie, now a high school graduate, left the farm soon after and joined the rest of their family in Chicago.

Hinda moved in with her son, Meyer, and his wife, Eleanor, who kept kosher. Si moved in with another sister who lived at 1862 Springfield on the more upscale side of the west-side Jewish ghetto. After spending six hundred dollars of his scant resources, his entire savings account, acquiring books and paying tuition, he enrolled at Crane College. The books, especially costly, turned out to be a poor investment. He found his education at South Haven High School entirely inadequate preparation for college, and he left Crane College after six weeks. He was forced to try his hand in Chicago's work force.

Si found part-time work in a sheet metal foundry that already employed one of his brothers. After working the graveyard shift, he returned home on the trolley bumping shoulders with the ill-humored, weary night shift workers and the ornery half-asleep dayshift workers anticipating the same sweat-stained collars. Weaving among them, foul-smelling, jobless drunkards and boozing washer-women, raucous in the half-asleep trolley, spewed putrid aromas toward Si's rural sensibilities. For Si, life in Chicago was too much of an eye-opener. It caused him to recoil like a hibernating bear that sticks its nose out in the midst of a vicious March storm. He longed for the innocence in the face of his friend, Kenny Dawe, the unobstructed view from horizon to horizon, preferring the sweet pungency of fresh manure to the assault of seedy, city odors and auto exhaust. He wanted something familiar, even if it meant leaving his family in Chicago.

Si went home to South Haven in 1930. With no place to stay, he turned to the man who had always treated the Rezniks like family. He had shared his timber across the road and had stolen into the Reznik kitchen every Christmas eve and hung stuffed stockings for each of the children. Bill Trowbridge, as the repossessed Reznik farm remained vacant, permitted Si to stay in the winter house. In this familiar place, Si started the second half of his life, the longer and more lucrative half. He found work as a deliveryman in the butcher shop in town. Already familiar with butchering farm animals, a necessity growing up in a strictly kosher farm household, he quickly learned the operation of the butchering business. Not only did Si inherit a comfort with schecting animals from his father, but also he shared his drive and integrity. He scaled the ladder from delivery boy to butcher and then to owner of a slaughterhouse and butcher shop. Si's path did not require his father's arduous climb from a conscripted private in the Russian army to owner of a small farm-resort and founder of a small synagogue across the ocean in Michigan, United States. Yet, Si had found an occupation and a success in life that his father would have recognized as the fulfillment of his own labor and hope for his son.

Epilogue

Si narrated his story with pride, equanimity, and fondness, without leaving his audience uncomfortable with how hard life seemed at the time. What he remembered was the shiny penny he earned towards his bicycle, not the cramped space or guano smears on his face; what he emphasized was the humor and the ingenious necessity of urinating out the second story screen in the winter, not the frigid conditions of his room. Still, he did not want me to confuse his fondness for his past for an existence that took place on easy street. His story was one of admiration, not for someone who changed the world, but someone who completely changed his world. He cherished the beauty in the simple rhythms of hard labor and the perseverance required to overcome its obstacles and problems. He wished to give his remote past the credit it was due for giving him his moral fiber and backbone. He recognized that his childhood molded him with loving hands wielding a blunt chisel.

Si must have had a premonition that his time to share his story was shrinking. Had he not sought me out when he did, some of his stories might have sunk into obscurity as his ability to remember them left him. Just in time, before the final curtain call, before the sun went down, before the lights went out, before the anesthetic took effect, he left his gift for those who care to hear it. While he was completely lucid, with his stories from his childhood, he wound his way back to those people among whom he would soon find himself.

Narrating his early memories allowed him to go home again and reaffirm the dignity of his beginnings.

After our sessions ended, as his physician, I was less successful keeping his mind steady than I was in recording some of its most precious and worthy memories. His aging mind began to malfunction more and more after he broke his hip and recovered enough to celebrate his 90th birthday with all those who loved and respected him in the community. His nights became interminable with delusions and wakefulness, pathetically controlled with today's medications. His days still had periods of clarity, but only a shadow of what they had been even two years earlier. There is no way he could have retrieved the gems from the deep pool of his memory murky in its rapid decline.

In broad brushstrokes, Si's memories leave a remarkable record of the dawn of electricity and the automobile, and of the difficulties of immigrating to a country of hope with little but a capacity to work hard. He enters images into the communal memory of the way things were done before industrialization. His narration stands as a monument to the heroes who work hard at the simple tasks that sustain life no less than a bronze statue stands for a war hero in the center of a plaza. I thank Si Reznik for transforming what might have been a burden into a privilege. He reveals the opportunity that exists for all of us to add to the communal memory and discover hidden heroes.